MEI structured mathematics
Differential Equations

JOHN BERRY
TED GRAHAM

Series Editor: Roger Porkess

MEI Structured Mathematics is supported by industry:
BNFL, Casio, ESSO, GEC, Intercity, JCB, Lucas, The National
Grid Company, Sharp, Texas Instruments, Thorn EMI

Hodder & Stoughton
A MEMBER OF THE HODDER HEADLINE GROUP

Acknowledgements

The authors and publishers would like to thank the following companies, institutions and individuals who have given permission to reproduce copyright material. The publishers will be happy to make suitable arrangements with any copyright holders whom it has not been possible to contact.

Illustrations were drawn by Jeff Edwards Illustration and Design.

Photographs:
J. Allan Cash (92 left).
Associated Press (92 right).
Life File/Emma Lee (2 top); Nicola Sutton (92 middle); Sue Wheat (130); Ron Williamson (19); Andrew Wood (112 right).
Mary Evans Picture Library (32 left); FF Bause (32 right); Lorgwo (56)
Ford (104)
Robert Harding Picture Library (19, 60)
Redferns/David Redfern (112 left)
Science Photo Library/Scott Camizine (146); John Greim (2 middle); Alfred Pasieka (26); Mark Phillips (2 bottom).

British Library Cataloguing in Publication Data

Berry, John, 1947 –
  Differential equations : Mechanics 4. – (MEI Structured Mathematics)
  1. Differential equations 2. Differential equaitons – Problems, exercises, etc.
  I. Title II. Graham, E.
  515.3 ' 5

ISBN 0 340 63085 X

First published 1996
Impression number 10 9 8 7 6 5 4 3 2 1
Year 2000 1999 1998 1997 1996

Typeset in Great Britain by Alden, Oxford, Didcot and Northampton.
Printed in Great Britain for Hodder & Stoughton Educational, a division of Hodder Headline Plc, 338 Euston Road, London NW1 3BH by Bath Press

MEI Structured Mathematics

Mathematics is not only a beautiful and exciting subject in its own right but also one that underpins many other branches of learning. It is consequently fundamental to the success of a modern economy.

MEI Structured Mathematics is designed to increase substantially the number of people taking the subject post-GCSE, by making it accessible, interesting and relevant to a wide range of students.

It is a credit accumulation scheme based on 45 hour components which may be taken individually or aggregated to give:

3 Components   AS Mathematics
6 Components   A Level Mathematics
9 Components   A Level Mathematics + AS Further Mathematics
12 Components A Level Mathematics + A Level Further Mathematics

Components may alternatively be combined to give other A or AS certifications (in Statistics, for example) or they may be used to obtain credit towards other types of qualification.

The course is examined by the Oxford and Cambridge Examinations and Assessment Council, with examinations held in January and June each year.

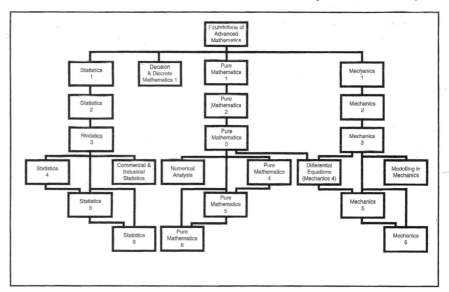

This is one of the series of books written to support the course. Its position within the whole scheme can be seen in the diagram above.

Mathematics in Education and Industry is a curriculum development body which aims to promote the links between Education and Industry in Mathematics at secondary school level, and to produce relevant examination and teaching syllabuses and support material. Since its foundation in the 1960s, MEI has provided syllabuses for GCSE (or O Level), Additional Mathematics and A Level.

For more information about MEI Structured Mathematics or other syllabuses and materials, write to MEI Office, 11 Market Place, Bradford-on-Avon, BA15 1LL.

Introduction

This book has been written primarily to support the MEI Structured Mathematics Component entitled Differential Equations (Mechanics 4). However it is also eminently suitable for use with any introductory course on differential equations in school, college or university.

The study of differential equations is important for applied mathematicians, scientists and engineers because these equations provide mathematical models for the many real world situations involving continuous change. Chapter 1 sets the scene by formulating differential equations for a wide range of situations. Subsequent chapters explore both theory and applications so that as well as learning how to do the mathematics you see its relevance, how to use it to express and solve real problems.

Chapters 2–5 explore the solutions of first order differential equations graphically, analytically and, in the case of chapter 5, numerically. Chapters 6–8 deal with the solutions of second order differential equations and their application to oscillating systems. In these chapters you will need an understanding of some basic concepts in mechanics, in particular kinematics and Newton's laws of motion.

Chapter 9 provides an introduction to the solution of systems of differential equations involving two or more dependent variables.

We would like to express our appreciation to the many people who have helped with advice and suggestions during the development of this book, in particular David Edsall, Mike Jones, John Reade, Judith Anstice and Diana Cowey. A number of MEI schools and colleges have piloted early drafts of the book, and we would also like to thank those teachers and students for the feedback they provided. Finally a special thanks to Karen Eccles for her patience and understanding in typing and retyping the many versions of the manuscript.

We would also like to thank those examination boards which have given permission for their past questions to be used in the exercises.

John Berry and Ted Graham

Contents

Using differential equations in modelling

It isn't that they can't see the solution.
It is that they can't see the problem.

G.K. Chesterton

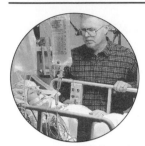

How does the quantity of a drug in the body vary with time?

How long does a cup of coffee take to cool?

After how many days will the moon take this shape again?

Each of these problems involves quantities that are continuously changing. If you think for a moment you will realise that there are many situations in which this is the case. We observe change in everything we see. Some of these changes are permanent: the coffee will cool to the temperature of the surroundings and then stay at that temperature. Some changes are cyclic: the shape of the moon in the sky follows a cyclic pattern.

The area of mathematics that describes quantities that change continuously is called *calculus*. The rate of change of a quantity is called its *derivative*, and equations that involve the derivative are called *differential equations*.

At its simplest, a differential equation provides information about the rate of change of one variable with respect to another in terms of those variables, for example

$$\frac{dy}{dx} = x + y, \qquad \frac{dP}{dt} = 0.02P$$

The *solution* of a differential equation provides a relationship between the variables themselves and does not involve derivatives. For these two equations the solutions are $y = -x - 1 + ce^x$ and $P = P_0 e^{0.02t}$ (where c and P_0 are constants). More complicated differential equations may involve higher derivatives e.g. $\dfrac{d^2 s}{dt^2}, \dfrac{d^3 y}{dx^3},$ or further variables.

Modelling with differential equations

The process of using mathematics to solve problems in the real world is called *mathematical modelling*. It is described by the flowchart in figure 1.1.

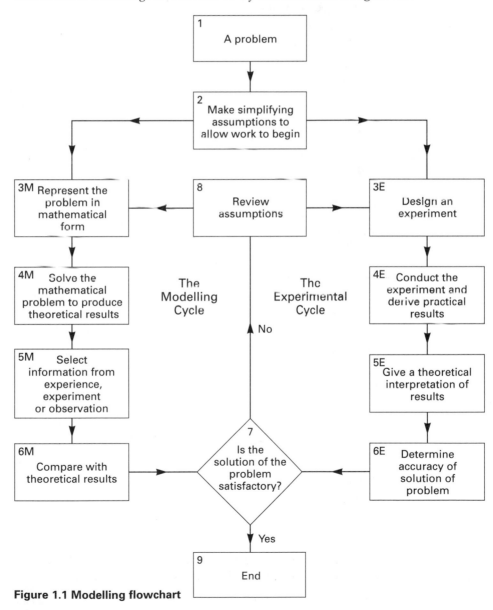

Figure 1.1 Modelling flowchart

In order to solve a problem successfully you need to find a mathematical model that describes the situation adequately. This model could take the form of an inequality, an equation or an algorithm (a set of steps). In this book you will meet problems for which the most appropriate model is a differential equation.

With many problems you will be able to go straight into the mathematics (the left hand cycle of the flowchart). This will usually be the case if the problem can be solved by applying models which are familiar to you. Most problems in mechanics are of this sort, because their solution involves the use of well-established models, such as Newton's laws of motion. Your solution to a problem will depend on the assumptions and simplifications you have made, so it may not be correct even if your calculations are sound. Even when you are solving a problem using modelling rather than experimental methods, you need some data to check your answer.

On the other hand, if no model is available, the first work you do after making your assumptions will be experimental (the right hand side of the flowchart). This should help you to understand the problem better, give you some figures to work with and help you to formulate a mathematical model.

To see how the modelling cycle works, we shall use it to analyse a medical problem. Caffeine is a stimulatory drug present in many drinks including coffee, tea and cola. It is known that too much caffeine in the body can be a health hazard. Suppose you are asked to predict the amount of caffeine remaining in a person's body some given interval after drinking a strong cup of coffee.

The first step is to understand the physiological processes involved in the situation, and to make simplifying assumptions to allow you to begin work on the problem (box 2 of the flowchart). The process of removing a drug from the human body varies with the drug. It involves a variety of organs, the most important of which are the kidneys. Removal of a drug by the kidneys (called 'renal clearance') releases the drug via the bladder into the urine. The rate of renal clearance of a drug can be measured by testing samples of urine.

You can begin by making the following simplifications and assumptions.

1. Once the coffee has been drunk, the caffeine is instantly absorbed into the bloodstream.
2. The caffeine is cleared from the bloodstream by renal clearance.
3. The rate of renal clearance is proportional to the quantity of the drug in the body.

Now you can formulate a mathematical model. If you let t represent time (in hours) and q represent the quantity (in mg) of caffeine present in the body tissues, statement **3** above can be written as:

$$\frac{dq}{dt} \propto q$$

k is a positive constant, the constant of proportionality.

or

$$\frac{dq}{dt} = -kq.$$

The negative sign is introduced because we know that the quantity of caffeine decreases with time.

This is an example of a differential equation. You will see later that its solution is given by

$$q = Ae^{-kt},$$

where A is a constant, the constant of integration. Although you could not yet have arrived at this solution for yourself, you can check that it is a solution of the differential equation by differentiating both sides:

$$\frac{dq}{dt} = A(-ke^{-kt}) = -kq.$$

So if $q = Ae^{-kt}$, then q satisfies the differential equation $\dfrac{dq}{dt} = -kq$.

The next step is to use information from experiments to find values of A and k. Suppose you have the following experimental results.

1. A cup of strong coffee contains 150 mg of caffeine.
2. The amount of caffeine cleared from the body in any one hour period is approximately 25% of the amount present at the start of the hour.

Substituting the data from **1** into our equation for q:
when $t = 0$, $q = 150$, so

$$150 = Ae^0 = A.$$

So in this case the constant of integration, A, is 150. Our equation for q is now

$$q = 150e^{-kt}.$$

Substituting the data from **2** into this equation:

when $t = 1$, $q = \dfrac{75}{100} \times 150 = 112.5$, so

$$112.5 = 150e^{-k}.$$

Dividing both sides by 150 and then taking logarithms gives the constant of proportionality as

$$k = -\ln\left(\frac{112.5}{150}\right) = 0.288 \quad \text{(correct to 3 significant figures)}$$

Based on the assumptions, simplifications and given data, the amount of caffeine at time t is given by
$$q = 150e^{-0.288t}.$$

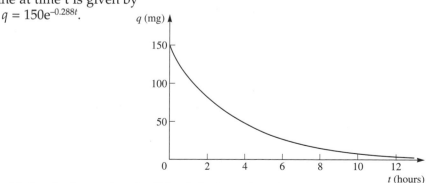

Figure 1.2 Graph of quantity of caffeine in body against time

This equation gives the amount of caffeine in the body as a function of time. It is a mathematical model formulated using both the modelling and experimental cycles. Further experimental information, such as tests of caffeine levels against time, would be needed to validate the model (box 6E of the modelling flowchart). For any real problem, you should expect to go round one or both cycles of the chart at least twice and possibly several times before you home in on a model that not only looks good on paper but actually matches the real world.

Activity

Use the model derived in the text, $q = 150e^{-0.288t}$, to help you answer these questions.

1. Calculate the approximate time taken for 50% of the caffeine to be eliminated from the body.
2. According to this model, after one cup of strong coffee the amount of caffeine in the body never again reaches zero. Do you think this is realistic?
3. Investigate the possibility of receiving a convulsive dose (10 g) of caffeine by drinking strong coffee, one cup at a time, at equal time intervals. (The convulsive dose is the dose which can cause death.)

For discussion

1. How is the rate of cooling of a cup of coffee related to its current temperature?
2. How is the rate of growth of a population related to its current size?
3. A chemical has spilled into a reservoir causing unsafe levels of pollution. Fresh water is fed into the reservoir from a river, and water is drained from the reservoir at the same rate. How does the concentration of the chemical in the reservoir change with time?

Forming differential equations from rates of change

If you are given sufficient information about the rate of change of a quantity, such as the caffeine level in the body or the height of water in a harbour, you can form a differential equation to model the situation. It is important to look carefully at the problem before writing down an equivalent mathematical statement. You have to decide whether you need a model for a rate of change with respect to time or with respect to another variable such as distance or height. You need to be familiar with the language used in these different cases.

If the altitude, h, of an aircraft is being considered, the phrase *rate of change of altitude* might be used. This actually means the *rate of change of altitude with respect to time*. You could write it as $\dfrac{dh}{dt}$ where t stands for time.

However, you might be more interested in how the altitude of the aircraft changes with the horizontal distance it has travelled. In this case you would talk about *the rate of change of altitude with respect to horizontal distance*, and you could write it as $\dfrac{dh}{dx}$ where x stands for the horizontal distance travelled.

You should be aware of two shorthand notations. Differentiation with respect to time is often indicated by writing a dot above the variable, for example $\dfrac{dx}{dt}$ may be written as $\dot{x}$, $\dfrac{d^2y}{dt^2}$ as $\ddot{y}$ (You say these as 'x dot' and 'y double dot'). Differentiation with respect to x may be denoted by the use of the symbol ', so that y' means $\dfrac{dy}{dx}$, f'' means $\dfrac{d^2f}{dx^2}$. (You say these as 'y dash' and 'f double dash'.)

Any equation which involves a derivative such as $\dfrac{dq}{dt}$, $\dfrac{dy}{dx}$ or $\dfrac{d^2x}{dt^2}$ is called a differential equation. A differential equation which involves a first derivative such as $\dfrac{dq}{dt}$ is called a *first order differential equation*. One which involves a second order derivative such as $\dfrac{d^2x}{dt^2}$ is called a *second order differential equation*. A third order differential equation involves a third order derivative, and so on.

The following examples show how differential equations can be formed from descriptions of a wide variety of situations.

SITUATION

The population of a country increases at a rate that is proportional to its present population. When the population is 50 million it is increasing at a rate of 2 million per year. Use this information to find a differential equation for the rate of change of the population.

Formulation

With variables t (for time in years) and P (for population in millions), the rate of increase of the population is written as $\dfrac{dP}{dt}$. The rate of increase is proportional to the current population, P, so you can write

$$\frac{dP}{dt} = kP$$

where k is the positive constant of proportionality.

You are also told that when $P = 50$, $\dfrac{dP}{dt} = 2$. You can substitute these values into the differential equation to find the value of k:

$$2 = k \times 50$$

$$k = \frac{1}{25}$$

So the differential equation is $\dfrac{\mathrm{d}P}{\mathrm{d}t} = \dfrac{P}{25}$.

> This is the rate of increase of population measured in millions per year.

SITUATION

Newton's law of cooling states that the rate at which a body cools is proportional to the difference between its temperature and the temperature of the surrounding air. Initially a kettle has a temperature of 100 °C and is cooling at a rate of 2 °C min $^{-1}$; the surrounding air has a temperature of 20 °C. Use this information to find an expression for $\dfrac{\mathrm{d}T}{\mathrm{d}t}$, where T is the temperature of the kettle in °C.

Formulation

According to Newton's law of cooling, the rate of cooling is proportional to the temperature difference between the kettle and the air. In this case you can write

$$\frac{\mathrm{d}T}{\mathrm{d}t} = -k(T-20)$$

> You know the temperature is decreasing. If you put a minus sign here you expect k to turn out positive. The minus sign could be omitted and in that case k would turn out negative.

When $T = 100$, $\dfrac{\mathrm{d}T}{\mathrm{d}t} = -2$, so

$$-2 = -k\,(100 - 20)$$

$$k = \frac{2}{80} = \frac{1}{40}.$$

The equation is therefore

$$\frac{\mathrm{d}T}{\mathrm{d}t} = -\frac{1}{40}(T - 20).$$

SITUATION

A simple model of the atmosphere above the Earth's surface is as a perfect gas at constant temperature. This means that

- the pressure, p, at any point is proportional to the density, ρ;
- the rate at which the pressure decreases with height, z, is proportional to the density.

Use this information to find an expression for $\dfrac{\mathrm{d}p}{\mathrm{d}z}$ in terms of p.

Formulation

Since the pressure, p, at any point is proportional to the density at that point, you can write

$$p = c_1\rho,$$

where c_1 is a positive constant.

Since the rate of change of pressure, p, with height, z, is proportional to the density, ρ, you can write

$$\frac{dp}{dz} = -c_2\rho,$$

Negative sign introduced because pressure decreases with height.

where c_2 is a positive constant.

Substituting for ρ from the first equation into the second gives

$$\frac{dp}{dz} = -\frac{c_2}{c_1}p = -cp$$

c is a constant formed from c_1 and c_2

SITUATION

A parachutist and her equipment have combined mass 70 kg and a terminal speed of 5 ms^{-1}. Using the modelling assumption that the air resistance R N is proportional to the speed v ms^{-1} of the parachutist, find an expression for $\frac{dv}{dt}$ at time t seconds. (Take g to be 10 ms^{-2}.)

Formulation

The diagram shows the forces acting on a parachutist of mass m. The force mg is the force of gravity acting vertically downwards, and R is the air resistance. You assume that the parachutist falls vertically in a straight line, and you take distances and speeds as positive in the downwards direction.

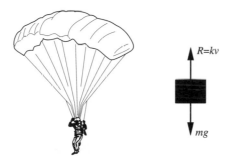

As well as using the information given, you need to use an established mathematical model, Newton's Second Law, to link the motion of the parachutist to the forces acting on her.

The resultant downward force is $mg - R$.

Applying Newton's Second Law, $F = ma$, at any instant gives

This side is F in $F = ma$

$$mg - R = m\frac{dv}{dt}$$

$\frac{dv}{dt}$ is the acceleration at that instant.

Dividing both sides by m gives

$$g - \frac{R}{m} = \frac{dv}{dt}$$

The modelling assumption that R is proportional to v gives $R = kv$ and so

$$\frac{dv}{dt} = g - \frac{kv}{m}.$$

This is the differential equation of motion for a parachutist of unknown mass. In this case the parachutist has mass 70 kg and reaches a terminal speed of 5 ms⁻¹, and you can use these values to calculate k.

At the terminal speed the acceleration is zero, i.e.

$$\frac{dv}{dt} = 0.$$

You can therefore write $g - \frac{kv}{m} = 0.$

Taking $g = 10$ and substituting $m = 70$ and $v = 5$ gives

$$10 - \frac{5k}{70} = 0.$$

so

$$k = \frac{700}{5} = 140.$$

Substituting for k in the expression for $\frac{dv}{dt}$ gives

$$\frac{dv}{dt} = 10 - \frac{140}{70}v = 10 - 2v.$$

This is the differential equation of motion for this particular parachute jump.

SITUATION

The volume, V, of a spherical raindrop of radius r is decreasing (due to evaporation) at a rate proportional to its surface area, S. Find an expression for $\frac{dr}{dt}$.

Formulation

Note that the problem description contains four variables, V, S, r and t, but that V and S are themselves functions of the radius r:

$$V = \frac{4}{3}\pi r^3 \quad \text{and} \quad S = 4\pi r^2.$$

You can use this to write the problem more simply in terms of the two variables, r and t.

The wording of the problem gives

$$\frac{dV}{dt} = -kS = -k(4\pi r^2).$$

where k is a positive constant. The next step is to write this equation in terms of r. The left hand side of the differential equation may be rewritten using the chain rule in the form

$$\frac{dV}{dt} = \frac{dV}{dr} \times \frac{dr}{dt}.$$

Since $V = \frac{4}{3}\pi r^3$, $\frac{dV}{dr} = 4\pi r^2$.

So $\frac{dV}{dt} = 4\pi r^2 \times \frac{dr}{dt}$.

The differential equation is now written

$$4\pi r^2 \frac{dr}{dt} = -4k\pi r^2$$

$$\Rightarrow \quad \frac{dr}{dt} = -k.$$

This very simple differential equation tells you that the radius decreases at a constant rate. You can solve this equation directly, by integration.

Exercise 1A

1. During the decay of a radioactive substance, the rate at which mass is lost is proportional to the mass present at that instant. Write down a differential equation to describe this relationship.

2. In an electrical circuit, the voltage is decreasing at a rate proportional to the square of the present voltage. When the voltage is 20 volts it is decreasing at a rate of 1 Vs^{-1}. Form a differential equation to describe the rate of change of voltage with time.

3. The rate at which the population of a particular country increases is proportional to its population. Currently the population is 68 million, and it is increasing at a rate of 2 million per year. Form a differential equation for P, the population of the country in millions.

4. The population of a colony of rabbits increases at a rate proportional to the population. When the population is 30 rabbits, it is increasing at a rate of 5 rabbits per month. Form a differential equation to model this situation.

5. The rate at which water leaves a tank is modelled as being proportional to the square root of the height of water in the tank. Initially the height of water is 100 cm and water is leaving at the rate of 20 cm^3 per second. Form a differential equation that describes this model.

6. A disease spreads so that the rate at which people are infected is proportional to the square root of the number already infected. When 100 people are known to be infected, new cases are appearing at a rate of 20 per day.
    Find an expression for $\frac{dN}{dt}$.

7. After a drug has been injected into the bloodstream, its concentration gradually decreases and the drug loses effectiveness. Assume that the decrease in concentration is proportional to the present concentration. Form a differential equation to model this situation, given that the concentration is decreasing by 0.02 gcm^{-3} s^{-1} when the concentration is 0.1 gcm^{-3}.

8. A hot object cools according to Newton's law of cooling. Initially it has a temperature of $80\,°C$ and is cooling at a rate of $0.05\,°Cs^{-1}$. The surrounding air has a temperature of $20\,°C$. Form a differential equation to describe the temperature loss.

9. The area of a circle is increasing at a rate proportional to its radius.

   (i) Write down an expression for $\dfrac{\mathrm{d}A}{\mathrm{d}t}$.

   (ii) Find an expression for $\dfrac{\mathrm{d}A}{\mathrm{d}r}$ in terms of r.

   (iii) Find an expression for $\dfrac{\mathrm{d}r}{\mathrm{d}t}$.

10. Air is escaping from a balloon at a rate proportional to its surface area. Given that the air is escaping at $4 \text{ cm}^3\text{s}^{-1}$ when the radius is 10 cm, find an expression for the rate of change of the radius, $\dfrac{\mathrm{d}r}{\mathrm{d}t}$.

    (Assume that the balloon is spherical.)

11. A poker of length 1m has one end in a fire. The temperature at each point of the poker remains the same at all times. However the temperature decreases along the length of the poker at a rate proportional to the distance from the hot end. A quarter of the way along the poker (from the hot end) the rate of decrease of the temperature is $16\,°\text{C}$ per centimetre. Formulate a differential equation to model this situation.

12. The water pressure in the sea increases with depth. At depth h below the surface, the rate of pressure increase with respect to depth is proportional to the density of sea water. The constant of proportionality is the acceleration due to gravity, g (9.8 ms^{-2}).

    The density ρ (in kg m^{-3}) of sea water in part of an ocean is modelled by

    $\rho = 1000\,(1 + 0.001h) \qquad 0 \le h \le 100$
    $\rho = 1100 \qquad\qquad\qquad h > 100$

    Find a differential equation to model this situation.

13. Heat-seeking missiles are designed to follow any object that emits heat. For example, an anti-aircraft missile can be 'locked on' to the heat of the exhaust from a jet engine, so that the missile always points towards the engine. This makes it harder for the aircraft to escape the missile by changing direction.

A heat-seeking missile is fired at an aircraft that is flying horizontally with constant speed v at a constant height of b. At the instant the missile is launched, the aircraft has co-ordinates (a, b) relative to the launch pad. At time t, the missile has position (x, y).

Formulate a differential equation for $\dfrac{\mathrm{d}y}{\mathrm{d}x}$, in terms of x, y and t, to model the path of the missile as it travels to intercept the aircraft.

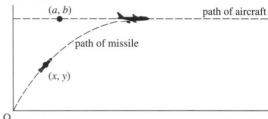

14. Water is pumped into a conical tank at a rate of 0.3 m^3 per second. The tank has the dimensions shown in the diagram, and the depth of the water is h metres at time t seconds.

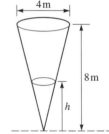

    Find an expression for $\dfrac{\mathrm{d}h}{\mathrm{d}t}$.

15. A volume V of water is held in a tank that has a square base, side 2 m, and height 4 m. Initially the tank is full. Water leaves the tank through a hole at the bottom at a rate of $\sqrt{20h}$ litres per second, where h is the depth of water in the tank. Find expressions for $\dfrac{\mathrm{d}V}{\mathrm{d}t}$ and $\dfrac{\mathrm{d}h}{\mathrm{d}t}$ in terms of h.

16. A can of drink is removed from a refrigerator, where it had a temperature of $2\,°\text{C}$, and placed in a warm room which is at a temperature of $22\,°\text{C}$. Form a differential equation that describes how the can warms up.

Constants of integration

The caffeine clearance problem (p4) is modelled by the first order differential equation

$$\frac{dq}{dt} = -kq,$$

and the constant of proportionality, k, has the value 0.288. Finding an expression for q is called *solving the differential equation*. In this case, the solution is

$$q = Ae^{-0.288t}.$$

This solution contains an unknown constant, A.

Some differential equations may be solved by direct integration. As you know, every time you integrate an expression, a constant of integration is produced. So when you solve a first order differential equation by direct integration you will end up with an expression that contains one constant of integration. If you solve a second order equation by direct integration, you end up with an expression that contains two constants of integration.

The same applies when you use other methods of solution: the solution of a first order differential equation contains one constant of integration, that of a second order equation contains two constants of integration, and so on.

The solution that contains these constants is called the *general solution* of the differential equation. Since it contains unknown constants, the general solution is actually a family of solutions, and it can be represented as a family of curves. For example, look at the differential equation

$$\frac{dy}{dx} = 2x - 1.$$

Integrating both sides with respect to x,

$$y = \int (2x - 1)\, dx$$
$$= x^2 - x + c$$

where c is the constant of integration.

> This is the general solution. Each value of c gives a different solution, and a different curve.

Some members of the family of solutions of this differential equation are shown in figure 1.3.

Suppose that you are now given one extra piece of information, such as 'when $y = 2$, $x = 3$'. You can use this information to find the particular value of c, and therefore the particular curve, that fits your problem.

Substituting $y = 2$ and $x = 3$ into the general solution gives

$$2 = 3^2 - 3 + c$$

and so $c = -4$.

The solution which fits your problem is

$$y = x^2 - x - 4.$$

This is a *particular solution* of the differential equation. It is represented in figure 1.3 by the curve which passes through $(0, -4)$. Now that you have identified a particular solution, you can find the value of y for any given value of x.

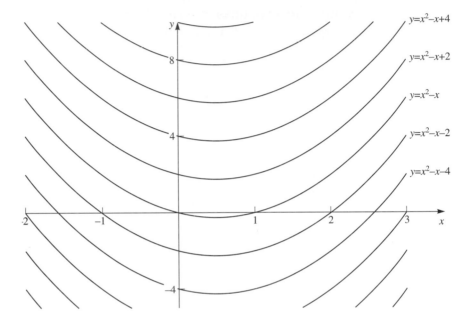

Figure 1.3 Some solutions of the differential equation $\dfrac{dy}{dx} = 2x - 1.$

The extra piece of information that allows you to identify the particular solution for your problem is called a *condition.* It effectively gives you the co-ordinates of one of the points through which the required solution curve passes. In the case of a second order differential equation, the general solution contains two constants of integration, and so two conditions are needed to find a particular solution. For a third order differential equation, three conditions are needed, and so on.

Verification of solutions

You will meet situations in which you think you know the solution of a differential equation without actually having to solve it. It may be that the situation, or one like it, is familiar to you, or you may have experimental data that suggest a solution. In such situations you need to check or *verify* that your solution does indeed satisfy the differential equation. You do this by substituting the solution into the differential equation, a process called *verification*. Verification is also helpful in cases where you have found a solution but just want to check your working.

EXAMPLE

Verify that $y = Ae^{-3x}$ is the general solution of the differential equation

$$\frac{dy}{dx} = -3y$$

where A is a constant. Sketch some members of the family of solutions.

Solution

Differentiating $y = Ae^{-3x}$ gives

$$\frac{dy}{dx} = -3Ae^{-3x} = -3(Ae^{-3x})$$

$$= -3y,$$

so $y = Ae^{-3x}$ is a solution of the differential equation $\dfrac{dy}{dx} = -3y$.

Since each value of A leads to a different solution curve, this is the general solution of the differential equation. Some members of the family of solutions are shown in the diagram. The curves all have equation $y = Ae^{-3x}$, but each has a different value of A.

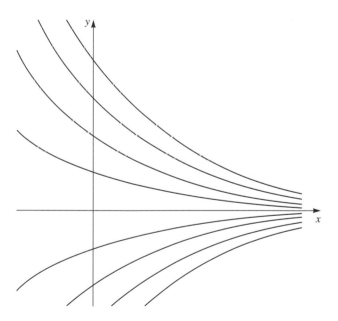

Note that if you are verifying a particular solution of a differential equation, you need to check

• that the expression satisfies the differential equation, and
• that it satisfies the given conditions.

EXAMPLE Verify that

$$v = 20e^{-2t} + 5$$

is the particular solution of the differential equation

$$\frac{dv}{dt} = 10 - 2v$$

that satisfies the condition $v = 25$ when $t = 0$.

Solution

Differentiating $v = 20e^{-2t} + 5$ gives

$$\frac{dv}{dt} = -40e^{-2t}.$$

The right hand side of the differential equation is $10 - 2v$. In this case,

$$
\begin{aligned}
10 - 2v &= 10 - 2(20e^{-2t} + 5) \\
&= 10 - 40e^{-2t} - 10 \\
&= -40e^{-2t} \\
&= \frac{dv}{dt}.
\end{aligned}
$$

> This is the given expression for v

So the solution $v = 20e^{-2t} + 5$ satisfies the differential equation

$$\frac{dv}{dt} = 10 - 2v.$$

Since the expression for v contains no unknown constants, it is a particular solution of the differential equation. We must now check that it satisfies the given condition, so we substitute $t = 0$ into the expression:

when $t = 0$, $v = 20e^0 + 5 = 25$ as required.

So $v = 20e^{-2t} + 5$ is the particular solution of the differential equation that satisfies the condition that $v = 25$ when $t = 0$.

Exercise 1B

1. Verify that $P = 40e^{t/5}$ is a solution of

$$\frac{dP}{dt} = \frac{P}{5}.$$

2. Verify that $M = 10e^{-5t}$ is a solution of

$$\frac{dM}{dt} = -5M.$$

3. (i) Verify that $v = 20(1 - e^{-0.5t})$ is a solution of the differential equation

$$\frac{dv}{dt} = 10 - 0.5v.$$

    (ii) What is the value of v when $t = 0$?

    (iii) What happens to v for large values of t?

4. The temperature, T, of an object changes such that

$$\frac{dT}{dt} = 2 - 0.1T.$$

    (i) Verify that $T = 20 + 60e^{-0.1t}$ is a solution of this differential equation.

    (ii) Verify that $T = 20 - 18e^{-0.1t}$ is also a solution.

    (iii) One of the solutions above models the temperature change of a cup of coffee placed in a room, and the other models that of a cold carton of juice just taken from a fridge. Which solution corresponds to which object?

5. A population grows so that

$$\frac{dP}{dt} = \frac{P}{20}.$$

(i) Verify that $P = Ce^{t/20}$ is the general solution of this equation.

(ii) When $t = 0$ the population is 50. Find the particular solution in this case.

(iii) Write down the particular solution corresponding to an initial population of 100.

(iv) Sketch a number of curves to illustrate the family of solutions.

6. A radioactive substance decays so that its mass m mg decreases according to the equation

$$\frac{dm}{dt} = -\frac{m}{100}.$$

(i) Verify that $m = 20e^{-t/100}$ is a solution of this differential equation.

(ii) Find the half-life (time taken for the mass to halve) for this substance.

(iii) What is the initial mass of the substance?

(iv) Write down the solution that would apply if the initial mass were 50 mg.

7. Verify that $y = \dfrac{-2}{x^2 + 2}$ is the solution of the differential equation

$$\frac{dy}{dx} = xy^2$$

that satisfies the initial condition $y = -1$ when $x = 0$.

8. By integrating both sides find the general solution of the differential equation

$$\frac{dy}{dx} = 3x^2 - x + 1.$$

Find the particular solution that satisfies the condition $y = 4$ when $x = 1$.

Sketch a number of curves to illustrate the family of solutions including the particular solution that passes through the point (1, 4).

9. Find the general solution of the differential equation

$$\frac{ds}{dt} = 4 - 10t.$$

Find the particular solution that satisfies the initial condition $s = 11$ when $t = 0$.

10. Which of the following are possible solutions of the differential equation

$$\frac{dy}{dx} = -8y?$$

A $\quad y = 4e^{-8x}$ B $\quad y = 8e^{-4x}$
C $\quad y = 4e^{-8x} + 2$ D $\quad y = 4e^{-8x} + 8$
E $\quad y = 8e^{-8x}$

11. (i) Verify that $P = P_0 e^{k(t - t_0)}$ is a solution of the differential equation

$$\frac{dP}{dt} = kP.$$

(ii) Verify that $P = Ae^{kt}$ is also a solution of the differential equation.

(iii) Express A in terms of k, t_0 and P_0.

12. (i) Verify that $T = \alpha + Ae^{-kt}$ is a solution of the differential equation

$$\frac{dT}{dt} = -k(T - \alpha) \qquad (k > 0)$$

which models Newton's law of cooling.

(ii) Find, in terms of A and α, the initial and final temperatures of the object.

(iii) An object at a temperature of 90 °C is placed in a room at 25 °C. State the values of α and A in this case.

(iv) Sketch this family of curves including cases when A is both positive and negative.

Exercise 1B continued

13. The acceleration of a stone sliding down a slope submerged in a fluid can be modelled as

$$\frac{dv}{dt} = 4 - v.$$

(i) Show that $v = Ae^{-t} + 4$ satisfies the differential equation.

(ii) Determine the value of A if the initial speed of the stone is 8 ms^{-1}.

14. Show that $y = Ae^x - (x^2 + 2x + 2)$ is a solution of the differential equation

$$\frac{dy}{dx} = y + x^2$$

15. In a model (which applies up to heights of about 10 km) for the pressure p in the lower atmosphere at a height z metres above sea level, it is assumed that gravity is constant and that the air behaves as a perfect gas. This gives the differential equation

$$\frac{dp}{dz} = -\frac{pg}{RT}$$

where g is the acceleration due to gravity, R is the universal gas constant and T is the absolute temperature of the air in Kelvin (K). The value of R is 8.314 JK^{-1} mol^{-1}.

The temperature distribution in the lower atmosphere is given by

$$T = 300 - 0.006z$$

(i) Taking g to be 10 ms^{-2}, show that this differential equation may, to good approximation, be written

$$\frac{1}{p}\frac{dp}{dz} = -\frac{200}{(50\,000 - z)}$$

(ii) Show that

$$p = p_0\left(1 - \frac{z}{50\,000}\right)^{200}$$

satisfies this differential equation, given that $p = p_0$ when $z = 0$.

KEY POINTS

- A differential equation is an equation involving derivatives such as

$$\frac{dy}{dx},\ \frac{dV}{dt}\ \text{or}\ \frac{d^2x}{dt^2}.$$

- The order of a differential equation is the order of its highest derivative.
- Differential equations are used to model situations which involve rates of change.
- The solution of a differential equation gives the relationship between the variables themselves, not their derivatives.
- The general solution of a first order differential equation satisfies the differential equation and has one constant of integration left in the solution.
- A particular solution of a differential equation is one in which additional information has been used to calculate the constant of integration.
- The general solution may be represented by a family of curves and a particular solution is one member of that family.
- To verify that a function is a particular solution you must check that the function satisfies both the given differential equation and the initial conditions.

Differential equations

2 Tangent fields

He . . . flung himself upon his horse and rode off madly in all directions.

Stephen Leacock

The photograph shows a shoal of fish. Shoals of fish can behave as if they were a single organism, displaying an ever-changing pattern of curves. At any instant, the body of each fish forms the tangent to one of the curves in the pattern.

In Chapter 1 you saw the solutions of differential equations represented graphically. The graphs were drawn using the algebraic, or *analytical*, form of the solutions. For many differential equations you cannot find an analytical solution. But even in these cases you can still predict the behaviour of the variables using graphical techniques, and this chapter shows you how to do so.

A differential equation gives you information about the rate of change of the dependent variable but not about the actual value of the variable at any instant. The information about the rate of change can be represented graphically as a *tangent field* (sometimes called a *direction field*), and this can be used to sketch approximate solution curves for the differential equation.

To see how this works, look again at the caffeine clearance problem of Chapter 1. The differential equation modelling the situation is

$$\frac{\mathrm{d}q}{\mathrm{d}t} = -0.288q.$$

For any value of q you can use this to calculate $\dfrac{\mathrm{d}q}{\mathrm{d}t}$. For example, if $q = 150$,

$\dfrac{dq}{dt} = -43.2$. So you know that when q has the value 150, whatever the value of t, the gradient of the solution curve is -43.2. You can represent this on a graph of q against t by drawing short dashes with gradient -43.2 at $q = 150$. (When working out the angle at which the lines should be drawn you will need to take into account the scales on your axes.) You can repeat this process for other values of q, until you have built up a tangent field for the differential equation, as in figure 2.1.

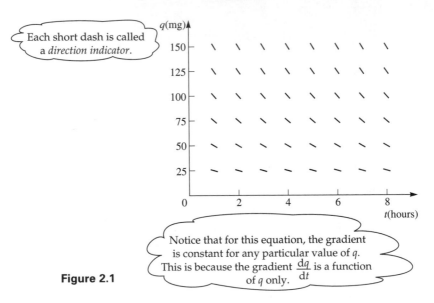

Each short dash is called a *direction indicator*.

Notice that for this equation, the gradient is constant for any particular value of q. This is because the gradient $\dfrac{dq}{dt}$ is a function of q only.

Figure 2.1

As you have seen, the general solution of a first order differential equation can be represented by a family of curves. Each curve represents one particular solution. For example, in the caffeine clearance problem you were interested in the particular solution corresponding to $q = 150$ when $t = 0$. Using the tangent field, you could sketch this curve, starting at the point (0, 150) and using the direction indicators as a guide. This has been done in figure 2.2. Curves have been sketched also for two different initial caffeine concentrations, $q = 100$ and $q = 50$. These represent two other particular solutions of the differential equation. You can see that any particular solution of the differential equation could be sketched in this way.

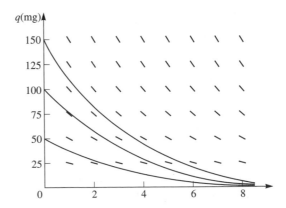

Figure 2.2

For the caffeine clearance problem you know the analytical solution:

$$q = Ae^{-2.88t}.$$

This means that you can calculate an accurate value for q at any time t. However, when an analytical solution is not available, or has a complicated form, you can use a tangent field to find approximate values, and to gain an understanding of the behaviour of the variable(s) in question.

EXAMPLE

For the differential equation

$$\frac{dy}{dx} = x^2 + y^2,$$

(i) set up a table of values giving $\dfrac{dy}{dx}$ for integer values of x and y, for

$-2 \leq x \leq 2$ and $-2 \leq y \leq 2$;

(ii) use the values in your table to draw the tangent field in the region covered by the table;

(iii) sketch the solution curve which passes through the point $(0, 1)$.

Solution

(i)

y \ x	−2	−1	0	1	2
2	8	5	4	5	8
1	5	2	1	2	5
0	4	1	0	1	4
−1	5	2	1	2	5
−2	8	5	4	5	8

(ii) and (iii)

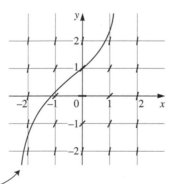

The gradient is always positive or zero. This is because $x^2 + y^2 \geq 0$

Notice that the gradient for a given value of y varies according to the value of x in this case. This is because the gradient is a function of both x and y.

Sketch of solution curve passing through $(0, 1)$ as required in part (iii).

Drawing tangent fields by hand can be very time-consuming. Many computer packages include a facility for drawing them, and this saves time; some of the tangent fields in this book are computer-generated. It is often useful, though, to look at the curves along which the direction indicators have the same gradient. These curves are called the *isoclines*, and they can be useful whether you are drawing the tangent field by hand or interpreting a computer display.

For the differential equation $\dfrac{dy}{dx} = 2y$, $\dfrac{dy}{dx}$ has the same value, namely 8, at the points $(-2,4)$, $(0,4)$, $(1,4)$, $(4,4)$. The line $y = 4$ is an isocline. In this example, all the lines parallel to the x axis are isoclines (figure 2.3).

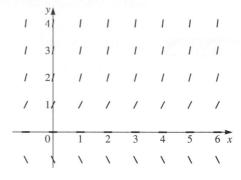

Figure 2.3 Isoclines for $\dfrac{dy}{dx} = 2y$

For the differential equation in the previous example, $\dfrac{dy}{dx} = x^2 + y^2$, the isocline for points where $\dfrac{dy}{dx}$ has value m is the circle $x^2 + y^2 = m$. This has centre the origin and radius $\sqrt{m}$ (figure 2.4).

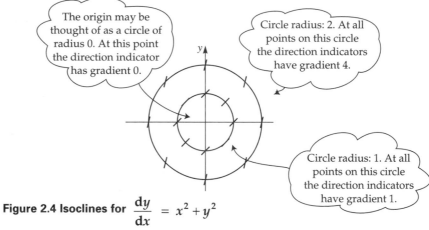

The origin may be thought of as a circle of radius 0. At this point the direction indicator has gradient 0.

Circle radius: 2. At all points on this circle the direction indicators have gradient 4.

Circle radius: 1. At all points on this circle the direction indicators have gradient 1.

Figure 2.4 Isoclines for $\dfrac{dy}{dx} = x^2 + y^2$

Information from the isoclines can often save a lot of work in drawing the tangent field. Having drawn the isoclines, it is then a simple matter to draw in as many direction indicators as you need to make the family of curves become clear.

Notice that for the differential equation $\dfrac{dy}{dx} = x^2 + y^2$, no analytical

solution is possible. The use of tangent fields has allowed us nonetheless to learn a great deal about the solution curves. Since many differential equations do not have analytical solutions, tangent fields provide a useful and powerful method.

In the next example the differential equation is one which can be solved analytically, but drawing the tangent field allows you to deduce the main features of the situation which it is modelling, without finding the analytical solution.

EXAMPLE

The differential equation

$$\frac{dv}{dt} = 10 - 5v$$

models the velocity of an object falling through a liquid.

(i) Sketch the tangent field for $0 \le t \le 2$ and $0 \le v \le 4$.

(ii) Superimpose the solution curves which represent an object that is initially

   (a) at rest;

   (b) moving at 4 ms^{-1}.

(iii) From the tangent field deduce the terminal speed of the object.

Solution

(i) The table below gives the values of $\frac{dv}{dt}$ for different values of v. (Note that in this case $\frac{dv}{dt}$ depends only on v and not on t.)

v	0	1	2	3	4
$\dfrac{dv}{dt}$	10	5	0	−5	−10

The values of the gradient $\frac{dv}{dt}$ become direction indicators in the tangent field. These are plotted in diagram A.

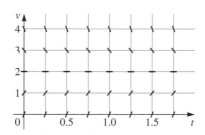

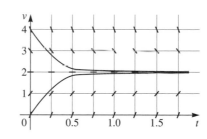

A   **B**

(ii) To superimpose a solution curve, you select the starting point and draw a curve from the point such that its gradient is always the same as that shown by the direction indicators. Diagram B shows the curves for initial speeds of 0 and 4 ms^{-1}.

(iii) The tangent field shows that as t increases all the solution curves approach the horizontal line $v = 2$. When $v = 2$, $\frac{dv}{dt}$ becomes zero and the speed does not change, so $v = 2$ is the terminal speed. (Notice how quickly the terminal speed is approached in both of the cases illustrated.)

1. (i) Copy and complete the table below, giving the values of $\dfrac{dy}{dx}$ where

$$\dfrac{dy}{dx} = 2x.$$

x y	−2	−1	0	1	2
2					
1					
0					
−1					
−2					

(ii) Sketch the tangent field.

(iii) Sketch the particular solutions that pass through (0, −2), (1,0) and (0,0).

(iv) Find the general solution of this differential equation analytically.

(v) Find the equations of the three curves you drew in part (iii).

(vi) Describe the isoclines for the tangent field you drew in part (ii).

2. (i) Construct a table for integer values of x and y from −2 to 2 giving the values of $\dfrac{dy}{dx}$ where

$$\dfrac{dy}{dx} = \dfrac{1-x}{y}.$$

(ii) Sketch the tangent field.

(iii) Superimpose particular solutions that pass through (0,1), (−1,0) and (0,0).

(iv) Find the equation of the line joining the points at which the gradient is $\frac{1}{2}$.

(v) Describe the isoclines for the tangent field.

3. Given that $\dfrac{dy}{dx} = y$,

(i) describe the isoclines of the corresponding tangent field;

(ii) draw the tangent field for $-3 \le x \le 3$ and $-3 \le y \le 3$;

(iii) sketch the curves which pass through (1, 2) and (3,−3).

4. Given that $\dfrac{dy}{dx} = 4 - y$,

(i) describe the isoclines and give the equation of the isocline joining points at which the gradient is 3;

(ii) draw the tangent field for $0 \le x \le 5$, $0 \le y \le 8$;

(iii) sketch in the curves which pass through (1, 2) and (1, 6).

(iv) What do you notice about the two curves you have just sketched?

5. (i) Given that $\dfrac{dy}{dx} = 4 - 2y$, draw a tangent field for $0 \le x \le 4, 0 \le y \le 4$.

(ii) Superimpose solution curves that pass through (0,4) and (0,1).

6. (i) Copy and complete the diagram by marking in accurately the direction indicators at the points marked with dots, given that

$$\dfrac{dy}{dx} = \dfrac{y}{x}$$

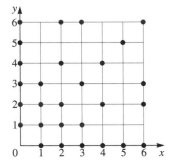

(ii) Give the equation of the isocline joining points at which the gradient is m (in the form $y = ...$).

(iii) Sketch in the curves which pass through the points
(a) (5, 5)   (b) (4, 0)   (c) (0, 3)
(d) (6, 3)   (e) (2, 6).

(iv) What do you notice about the five curves that you have just sketched? Use this information to draw the curve through the point (4,6).

(v) Why can't you mark in the direction indicator at the origin?

7. Given that $\dfrac{dy}{dx} = x + y,$

(i) state the equation of the isocline joining points at which the gradient is m;

(ii) draw the tangent field for $-2 \le x \le 2$, $-2 \le y \le 2$;

(iii) on your diagram, draw the isocline which joins points at which the gradient is zero;

(iv) sketch in the curves which pass through (a) $(0, -1)$ (b) $(2,2)$.

8. Given that $\dfrac{dy}{dx} = y - x,$

(i) state the equation of the isocline joining points at which the gradient is m;

(ii) draw the tangent field for $-2 \le x \le 2$, $-2 \le y \le 2$;

(iii) on your diagram draw the isocline which joins points at which the gradient is zero;

(iv) sketch in the curves which pass through (a) $(1, 2)$ (b) $(2, -1)$.

9. Given that $\dfrac{dy}{dx} = xy,$

(i) state the equation of the isocline joining points at which the gradient is m, and state the name given to this curve;

(ii) draw the tangent field for $-2 \le x \le 2$, $-2 \le y \le 2$;

(iii) describe the isocline which joins points at which the gradient is zero;

(iv) sketch in the curves which pass through

(a) $(2, 2)$ (b) $(0, -1)$ (c) $(0, 0)$.

10. The rate of change of current in a circuit is modelled by

$$\frac{di}{dt} = 3 - i \,.$$

(i) Draw a tangent field for $0 \le i \le 4$ and $0 \le t \le 4$.

(ii) Draw the solution curve that represents an initial current of 0.

(iii) What is the current in the circuit after a long period of time?

11. The population, P, (in millions) of a country is increasing. The rate of increase is modelled by

$$\frac{dP}{dt} = \frac{P}{20}$$

where t is the time in years after 1 January 1990.

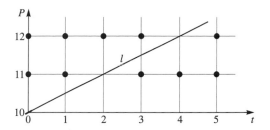

The direction indicator at the point where $t = 0$, $P = 10$ has already been drawn. The line l whose gradient is $\frac{1}{2}$ and which passes through the point where $t = 0$, $P = 10$ has also been drawn. Copy and complete the diagram by marking in as accurately as possible the direction indicators at the ten points marked with dots.

On 1 January 1990, the population was 10 million. Draw the curve which passes through this point, deciding carefully on which side of the line l the curve should lie.

When did the population increase to 12 million: was it during 1992, 1993 or 1994?

12. A parachutist falls such that

$$\frac{dv}{dt} = 8 - v$$

where v ms^{-1} is the speed at time t s.

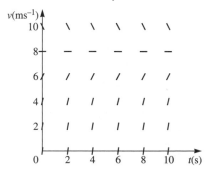

(i) The diagram shows the tangent field for $0 \leq v \leq 10$ and $0 \leq t \leq 10$.

(ii) Sketch solution curves corresponding to initial speeds

(a) 0 ms^{-1} (b) 5 ms^{-1} (c) 10 ms^{-1}.

(iii) What is the terminal speed in each of these cases?

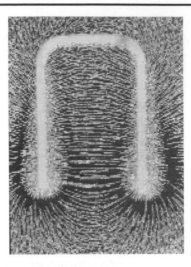

The curves in a magnetic field (shown by the iron filings in the photo), indicate the direction of the magnetic force at any point.

Investigation

(i) Display the tangent field for

$$\frac{\mathrm{d}y}{\mathrm{d}x} = -\frac{(\alpha + x)}{(\beta + y)}$$

using a variety of values for α and β.

Describe the solution curves for the differential equation in terms of α and β. What is the equation of the curves produced?

(ii) Now investigate the solution curves of

$$\frac{\mathrm{d}y}{\mathrm{d}x} = \frac{(\alpha + x)}{(\beta + y)}$$

Describe these curves as fully as possible.

KEY POINTS

- The tangent field is a diagram showing the direction indicators at a number of points.
- Drawing a tangent field allows you to see the main features of the solution of a differential equation even when you cannot find the solution analytically.
- An isocline is the locus of points at which the direction indicators have equal gradients.
- Isoclines are helpful when you are drawing or interpreting tangent fields.

3

Separation of variables

Here and elsewhere we shall not obtain the best insight into things until we actually see them growing from the beginning.

Aristotle

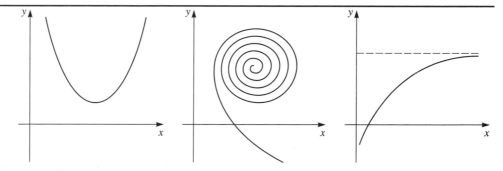

Look at the three curves drawn above. For each one, state whether it could be a particular solution of the differential equation

$$\frac{\mathrm{d}y}{\mathrm{d}x} = kx^n$$

where k is a constant, n a real number (positive or negative, integer or non-integer).

In Chapters 1 and 2 you have seen that differential equations arise in many mathematical problems, and you have met one approximate method of solving first order differential equations. In this chapter you meet a method of finding analytical solutions of some first order differential equations.

There is one class of differential equations that you can already solve analytically. Look at the equation

$$\frac{\mathrm{d}y}{\mathrm{d}x} = 2x^2$$

Direct integration gives

$$y = \int 2x^2 \, \mathrm{d}x = \tfrac{2}{3} x^3 + c.$$

c is a constant of integration

You could use direct integration because the differential equation was of the form

$$\frac{\mathrm{d}y}{\mathrm{d}x} = \mathrm{f}(x)$$

and you could integrate the function f(x).

Now look at the differential equation

$$\frac{\mathrm{d}y}{\mathrm{d}x} = xy^2 .$$

This equation cannot be solved by direct integration because the right hand side contains y as well as x. However, this particular equation is still relatively easy to solve, since the right hand side is a product of a function of x and a function of y. You can proceed as follows.

First you *separate the variables*:

$$\frac{1}{y^2}\frac{dy}{dx} = x$$

Notice that the right hand side is now a function of x only, and the left hand side is $\dfrac{dy}{dx}$ times a function of y only.

This can be written as

$$\int \frac{1}{y^2}\,dy = \int x\,dx$$

Now both sides can be integrated separately.

Integrating the left hand side with respect to y gives $\displaystyle\int\frac{1}{y^2}\,dy = -\frac{1}{y}+c_1$.

Integrating the right hand side with respect to x gives $\displaystyle\int x\,dx = \frac{1}{2}x^2 + c_2$.

You now have the equation

$$-\frac{1}{y}+c_1 = \frac{1}{2}x^2 + c_2$$

or

$$-\frac{1}{y} = \frac{1}{2}x^2 + c_2 - c_1.$$

This equation includes two arbitrary constants, c_1 and c_2. But since $c_2 - c_1$ is itself an arbitrary constant we would normally replace it with just one constant of integration, c, on the right hand side:

$$-\frac{1}{y} = \frac{1}{2}x^2 + c \qquad \text{or} \qquad y = -\frac{1}{\frac{1}{2}x^2 + c}.$$

This can be written more tidily as

$$y = -\frac{2}{x^2 + k}$$

where the new arbitrary constant $k = 2c$.

The differential equation $\dfrac{dy}{dx} = xy^2$ is said to be *separable* because it has the form

In the case above
$f(x) = x$
$g(y) = y^2$

$$\frac{dy}{dx} = f(x)g(y).$$

Such an equation can be rewritten in the form

$$\int \frac{1}{g(y)}\,dy = \int f(x)\,dx.$$

EXAMPLE Solve the differential equation $\dfrac{dy}{dx} = \dfrac{\sin x}{y^2}$

Solution

Multiplying both sides of the equation by y^2:

$$y^2 \frac{dy}{dx} = \sin x$$

This can be written as

$$\int y^2 \, dy = \int \sin x \, dx$$

> This process has separated the variables.

$$\tfrac{1}{3} y^3 = -\cos x + c$$

$$y = \sqrt[3]{-3\cos x + 3c} = \sqrt[3]{A - 3\cos x}$$

where $A = 3c$ is a constant. This is the general solution of the differential equation.

The same approach is used to solve equations of the form

$$\frac{dy}{dx} = g(y),$$

as in the following examples.

EXAMPLE The population, P (in millions), of a country grows so that

$$\frac{dP}{dt} = \frac{P}{50}$$

(i)   Find the general solution of this differential equation.
(ii) Find the particular solution if $P = 100$ when $t - 0$.

Solution

(i)   Dividing both sides of the equation by P gives

$$\frac{1}{P} \frac{dP}{dt} = \frac{1}{50}$$

$$\Rightarrow \int \frac{1}{P} dP = \int \frac{1}{50} dt$$

$$\ln |P| = \frac{t}{50} + c.$$

Since the population P is always positive, $|P| = P$.

Rearranging gives

$$P = e^{\left(\frac{t}{50} + c\right)}$$

which can be written as

$$P = Ae^{\frac{t}{50}}$$

where $A = e^c$.

So the general solution is

$$P = Ae^{\frac{t}{50}}.$$

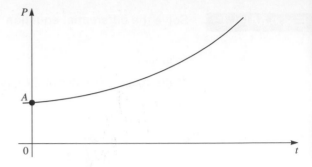

Notice that when $t = 0$, $P = A$. This is the initial population, often denoted by P_0.

(ii) To find the particular solution you substitute $P = 100$ and $t = 0$ into the general solution, to obtain

$$100 = Ae^0$$

$$\Rightarrow \quad A = 100.$$

Using this value of A gives the particular solution

$$P = 100e^{\frac{t}{50}}.$$

In the previous example, it was possible to drop the modulus sign because population is always positive: $|P| = P$. It is often not possible to drop the modulus sign, as in the next example.

EXAMPLE

A drink is placed in a room where the ambient temperature is 20 °C. A model for the subsequent temperature, T, of the drink at time t, in hours, is given by Newton's law of cooling. This leads to the differential equation

$$\frac{dT}{dt} = -5(T - 20) .$$

Find the solution of this differential equation, in the cases where
(i) $T = 80$ when $t = 0$;
(ii) $T = 0$ when $t = 0$.

Solution

Dividing both sides of the differential equation by $(T - 20)$, and rewriting it as two integrals:

$$\int \frac{1}{T-20}\, dT = \int -5\, dt \qquad \text{This is the process of separating the variables.}$$

$$\Rightarrow \quad \ln|T - 20| = -5t + c$$
$$|T - 20| = e^{(-5t+c)}$$

$$|T{-}20| = Ae^{-5t} \qquad \text{where } A = e^c.$$

This is the general solution of the differential equation. There are two different initial conditions to consider, and each of these will give a different particular solution.

(i)   Here $T > 20$, so $(T - 20)$ is positive and $|T - 20| = T - 20$.

The general solution can therefore be written as
$$T - 20 = Ae^{-5t}$$
or
$$T = 20 + Ae^{-5t}.$$

Substituting $T = 80$ and $t = 0$ gives
$$80 = 20 + Ae^0 = 20 + A$$
$$\Rightarrow \quad A = 60.$$
The particular solution in this case is
$$T = 20 + 60e^{-5t}.$$

The drink is cooling down to the ambient temperature of 20 °C.

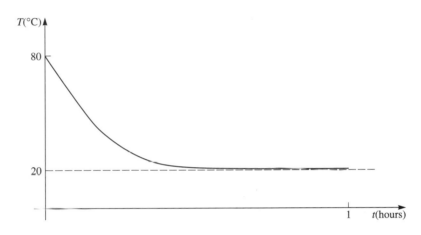

(ii)   Here $T < 20$, so $(T - 20)$ is negative and $|T - 20| = 20 - T$.

The general solution can therefore be written as
$$20 - T = Ae^{-5t}$$
or
$$T = 20 - Ae^{-5t}.$$

Substituting $T = 0$ and $t = 0$ gives
$$0 = 20 - Ae^0 = 20 - A$$
$$\Rightarrow \quad A = 20.$$
The particular solution is $T = 20 - 20e^{-5t}$.

The drink is warming up to the ambient temperature of 20 °C.

The discovery of calculus is usually attributed to Isaac Newton with the invention of the method of fluxions in 1665, when he was 23 years old, although it is also claimed that Gottfried Leibniz discovered it first but did not publicise his work. The study of differential equations followed on naturally. Newton is known to have solved a differential equation in 1676 but he only published details of it in 1693, the same year in which a differential equation featured in Leibniz's work.

The method of separation of variables was developed by Johann Bernoulli between 1694 and 1697. He was a famous teacher and wrote the first calculus textbook in 1696; one of his students was Leonhard Euler (see chapter 5). Johann Bernoulli was one of the older members of a quite remarkable family of Swiss mathematicians (three of them called Johann), spread over three generations at least eight of whom may be regarded as famous.

Johann Bernoulli

Gottfried Leibniz

Exercise 3A

1. Which of the following differential equations can be solved by the method of separation of variables? Give the general solutions of those which can be solved by this method.

(i) $\dfrac{dy}{dx} = y$

(ii) $\dfrac{dy}{dx} = xy$

(iii) $\dfrac{dy}{dx} = x^3 y$

(iv) $\dfrac{dy}{dx} = 3x^2 e^{-y}$

(v) $\dfrac{dy}{dx} = x + yx$

(vi) $\dfrac{dy}{dx} = x + y^2$

(vii) $\dfrac{dy}{dx} = e^{x+y}$

(viii) $\dfrac{dy}{dx} = \dfrac{y}{x(x-1)}$

(ix) $\dfrac{dy}{dx} = x^2 - y^2$

(x) $\dfrac{dy}{dx} = \dfrac{\sin x}{y^2}$

(xi) $\dfrac{dy}{dx} = \dfrac{y+2}{x-2}$

(xii) $\dfrac{dx}{dt} = x^2 - 8x$

2. Find the particular solution of each of the differential equations below for the given conditions:

   (i) $x\dfrac{dy}{dx} = y^2$; $y = 10$ when $x = 1$.

   (ii) $\dfrac{dy}{dx} = \dfrac{y^2}{x}$; $y = 2$ when $x = 1$.

   (iii) $\dfrac{dy}{dx} = \dfrac{x^2}{y}$; $y = 10$ when $x = 1$.

   (iv) $\dfrac{dy}{dx} = e^{-y}\sin 2x$; $y = 0$ when $x = 0$.

   (v) $\dfrac{dy}{dx} = x^2e^{-y}$; $y = 10$ when $x = 0$.

   (vi) $\dfrac{dy}{dx} = \dfrac{e^y}{y}$; $y = 2$ when $x = 0$.

3. The rate of radioactive decay of a chemical is given by

$$\frac{dm}{dt} = -5m.$$

   (i) Find the general solution of this equation.

   (ii) Given that $m = 10$ when $t = 0$, find the particular solution.

4. A bacterium reproduces such that the rate of increase of the population is given by:

$$\frac{dP}{dt} = 0.7P,$$

   where t is measured in minutes.

   (i) Find the general solution of this differential equation.

   (ii) Find the particular solution if $P = 100$ when $t = 0$.

   (iii) How long does it take for the population to double?

5. An object is projected horizontally on a surface with an initial speed of 20 ms^{-1}. Its speed is modelled by the differential equation

$$\frac{dv}{dt} = -\frac{v^2}{10}.$$

   (i) Find the particular solution of this differential equation that corresponds to the initial condition given.

   (ii) How long does it take for the speed to drop to 10% of its original value?

6. Water evaporates from a conical tank such that the rate of change of the height of water is modelled by the differential equation

$$\frac{dh}{dt} = -\frac{\pi h^2}{4}.$$

   (i) Find the general solution of this differential equation.

   (ii) Initially the height of the water is H. Find, in terms of H, how long it takes for the height of the water in the tank to decrease by 10%.

7. Water is leaving a tank through a pipe at the bottom, such that the change in the height of water is modelled by the differential equation

$$4\frac{dh}{dt} = -\sqrt{20h},$$

   where t is in minutes and h in centimetres.

   (i) Find the particular solution of this differential equation given that the initial height is 4 cm.

   (ii) How long does the tank take to empty?

8. The differential equation below models the motion of a body falling vertically subject to air resistance.

$$\frac{dv}{dt} = 10 - 0.2v$$

   where v is the downward vertical speed of the body in ms^{-1} and time, t, is measured in seconds.

   (i) Find the general solution of the differential equation.

   (ii) Find particular solutions when

       a) the body starts at rest

       b) the body has initial downwards vertical speed 80 ms^{-1}.

   In each case state the terminal velocity.

9. The temperature of a hot body is initially 100 °C. It is placed in a tank of water at a temperature of 20 °C. The rate of change of temperature is modelled by the differential equation

$$\frac{dT}{dt} = -0.5(T - 20),$$

where t is the time in minutes.

(i) Find the particular solution of this differential equation that fits the initial condition given.

(ii) After what time interval has the temperature of the body fallen to 50 °C?

10. A small particle moving in a fluid satisfies the differential equation

$$\frac{dv}{dt} = -0.2(v + v^2).$$

Find the particular solution of this differential equation given that $v = 40$ when $t = 0$.

11. For the electrical circuit shown below, the current, i amperes, flowing once the switch is closed is given by

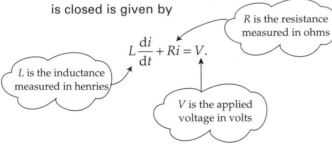

$$L\frac{di}{dt} + Ri = V.$$

R is the resistance measured in ohms

L is the inductance measured in henries

V is the applied voltage in volts

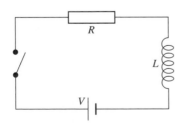

(i) If $L = 0.04$, $V = 20$ and $R = 100$, find the general solution of the differential equation.

(ii) If $i = 0$ when $t = 0$ find the particular solution.

(iii) Find the general solution of the original differential equation in terms of R, V and L.

12. A canal lock is modelled as a tank with a rectangular base of area 50 m². When the lock is being filled, the height of the water surface above its lowest level is h m at time t, where t is the number of seconds after the lock starts to fill.

In the model, h satisfies the differential equation

$$\frac{dh}{dt} = k(2.5 - h),$$

where k is constant.

When $t = 0$, water flows into the lock at a rate of 2 m³ s⁻¹.

(i) Show that initially $\frac{dh}{dt} = 0.04$ ms⁻¹, and hence that $k = 0.016$.

(ii) Solve the original differential equation for t in terms of h.

The lock is full when $h = 2.5$.

(iii) According to your solution, how long does it take to fill the lock?

(iv) How long would it take for the water level to rise to within 2 mm of the lock being full? (In practice, the lock gates can be opened before the lock is completely full.)

[MEI]

13. A car of mass 1000 kg is initially at rest, and is pushed by a man with a force F in newtons. The value of F decreases linearly with respect to the speed v ms⁻¹ of the car, as shown in the graph below. The man cannot push when running at a speed greater than 4 ms⁻¹.

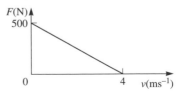

The car is on level ground, and resistances to motion may be neglected.

(i) Show that the motion of the car satisfies the differential equation

$$\frac{dv}{dt} = \frac{(4-v)}{8},$$

where t is the time in seconds for which the man has been pushing.

Write down the initial condition.

(ii) Solve the differential equation to show that $v = 4(1 - e^{-t/8})$.

(iii) Sketch the speed–time graph of the car's motion.

(iv) How long does it take the car to reach a speed of 3 ms^{-1}?

[MEI]

14. At 3.00 a.m. one morning the police were called to a house where the body of a murder victim had been found. The police doctor arrived at 3.45 a.m. and took the temperature of the body, which was 34.5 °C. One hour later he took the temperature again and measured it to be 33.9 °C. The temperature of the room was fairly constant at 15.5 °C.

Using Newton's law of cooling as a model, estimate the time of death of the victim. (The normal body temperature of a human being is 37.0 °C.)

15. A tank contains 2000 litres of salt solution with an initial concentration of 0.3 kg l^{-1}. It is necessary to reduce the concentration to 0.2 kg l^{-1}. In order to do this pure water is pumped into the tank at a rate of 8 litres per minute and the solution is pumped out of the tank at the same rate. The liquid in the tank is stirred so that perfect mixing may be assumed. At time t minutes the mass of salt in the tank is denoted by M kg and the concentration of salt in the solution by $C \text{ kg l}^{-1}$.

(i) Explain how this information leads to the results

$$\frac{dM}{dt} = -8C \text{ and } M = 2000C.$$

(ii) Hence form a differential equation for C.

(iii) After how long is the concentration reduced to 0.2 kg l^{-1}? Give your answer to the nearest minute.

16. A particle of mass m, travelling through air, experiences a resistive force proportional to its speed v. The resistance is such that in free fall it would reach a terminal speed V.

(i) Determine the value of the constant of proportionality.

The particle is now projected vertically upwards with a speed U.

(ii) Show that at time t after projection its speed v satisfies the differential equation

$$\frac{dv}{dt} = -\left(1 + \frac{v}{V}\right)g.$$

(iii) Using $\dfrac{dv}{dt} = v\dfrac{dv}{dx}$, where x is the height above the point of projection, show that the particle reaches a maximum height of H where

$$H = \frac{V^2}{g}\left\{\frac{U}{V} - \ln\left(1 + \frac{U}{V}\right)\right\}.$$

(iv) Show that if U is much smaller than V then

$$H \approx \frac{U^2}{2g}\left(1 - \frac{2U}{3V}\right).$$

Comment on the significance of this result in terms of the situation being modelled.

[MEI]

17. This problem concerns the growth of sweet peas in John Berry's garden in Devon in the summer of 1994. Two dozen plants were grown and their average heights were recorded every seven days. The average height when measurements began was 8 cm.

week	height (cm)
0	8
1	14
2	26
3	47
4	69
5	93
6	121
7	146
8	162
9	177
10	182

(i) An initial model assumes that the rate of growth of a plant is proportional to its height h. Formulate a differential equation for this model.

(ii) Solve the differential equation, using the initial condition and the average height after week 1 to find the two unknown constants in the solution. This is called the exponential model.

(iii) Plot the real data, and superimpose a graph of the exponential model. Criticize the exponential model. Give reasons why you think it does not fit the data for higher values of t.

(iv) A revised model is the *logistic differential equation*

$$\frac{\mathrm{d}h}{\mathrm{d}t} = ah\left(1 - \frac{h}{H}\right),$$

where H is the final ('equilibrium') height of the sweet pea plants and a is a constant. Without solving this equation, explain why it might be an improvement on the exponential model.

(v) Solve the logistic differential equation. Use the data in the table for weeks $0, 5$ and 10 to find the unknown constants in the solution.

(vi) Calculate the table of values for the logistic model for $t = 0$ to $t = 10$. Comment on the appropriateness of the logistic equation as a model.

18. A simple model of the progression of a virus is being formulated. A closed population of constant size 10^5 consists entirely of those infected by the virus (*infectives*) and those not yet infected (*susceptibles*). The rate of increase of infectives is proportional to the product of the number I of these infectives and the number S of susceptibles where $I + S = 10^5$. Time t is measured in weeks.

(i) Show that

$$\frac{\mathrm{d}I}{\mathrm{d}t} = kI(10^5 - I)$$

where k is a constant of proportionality and write down an equivalent equation for $\dfrac{\mathrm{d}S}{\mathrm{d}t}$ in terms of S.

Initially 1% of the population are infectives and it is noted that 10 weeks later this has risen to 20%.

(ii) Show that to three significant figures $k = 3.21 \times 10^{-6}$.

(iii) Find I as a function of time and hence find S as a function of time.

(iv) Verify that your expression for S in part (iii) satisfies the differential equation in part (i) and the initial conditions.

(v) Determine the time at which the number of infectives equals the number of susceptibles.

[MEI]

Investigations

Coffee Cooling

Newton's law of cooling says that in well ventilated conditions, the rate of cooling of an object is proportional to the temperature difference between the object and its surroundings. Investigate whether this is an appropriate model for a cup of hot coffee in a room with the door and windows closed. If not, use your data to suggest a better model.

Fish stocks

The fish in oceans, rivers and lakes constitute a major food source. However with the introduction of modern fishing methods there is the danger of some species being overfished to the point where their numbers are seriously reduced; in extreme cases there is risk of extinction.

In this investigation you are to consider the effect of fishing on a single species at time t in a limited environment. The population of the species at time t is P, and the maximum the environment can sustain is P_m.

The rate of change of the population is given by

$$\frac{dP}{dt} = I - D.$$

The variable D is the rate of decrease caused by fishing, and it depends upon the fishing strategy. The variable I is the rate of natural increase, given by

$$I = kP(P_m - P) \quad \text{where } k \text{ is a constant.}$$

This is called the logistic model.

Investigate the following two cases

(i) Constant catch: in this case $D = c$, a constant.
(ii) Periodic harvesting: in this case $D = 0$, but at regular intervals, P is reduced by a fixed amount.

KEY POINTS

- The method of separation of variables can be used to solve first order differential equations of the form

$$\frac{dy}{dx} = f(x)\,g(y).$$

- Separating the variables gives

$$\int \frac{1}{g(y)}\,dy = \int f(x)\,dx$$

Integrating factors

There is no branch of mathematics, however abstract, which may not some day be applied to phenomena of the real world.

Nikolai Lobatchevsky

Which of the following differential equations can be written in the form $\dfrac{dy}{dx} = f(x)g(y)$ and which cannot?

(i) $\dfrac{dy}{dx} = x^2 + x^2 y$

(ii) $\dfrac{dy}{dx} = x^2 - xy$

(iii) $\dfrac{dy}{dx} = x^2 + x + xy + y$

(iv) $\dfrac{dy}{dx} = x + y^2$

When a first order differential equation cannot be written in the form $\dfrac{dy}{dx} = f(x)g(y)$, it cannot be solved using the method of separation of variables. If however it is a *linear* equation, you can multiply it by a special function called an *integrating factor* which converts it to a form which can be integrated. This method is the subject of this chapter.

Linear equations

A linear differential equation is one in which the independent variable (y in these examples) only appears to the power of 1. So the differential equation $\dfrac{dy}{dx} = x^2 - xy$ is linear because the only terms that involve y are $\dfrac{dy}{dx}$ and $-xy$. There are no terms in y^2, $\sqrt{y}$, $\sin y$, $\left(\dfrac{dy}{dx}\right)^2$ etc. (Strictly it is linear in y.)

Any linear first order differential equation may be written in the form

$$\frac{dy}{dx} + Py = Q$$

where P and Q are functions of x only.

For example, the equation

$$\frac{dy}{dx} = x^2 - xy$$

can be rewritten as

$$\frac{dy}{dx} + xy = x^2.$$

This is in the form

$$\frac{dy}{dx} + Py = Q,$$

where the functions P and Q are given by $P = x$ and $Q = x^2$.

For discussion

Which of the four equations given at the start of the chapter are linear?

Identify the functions P and Q for each of those equations.

Look at the next two examples which concern linear differential equations that can be solved by close inspection.

EXAMPLE

Find the general solution of the differential equation

$$(\cos x)\frac{dy}{dx} - (\sin x)y = x^2$$

Solution

The equation looks forbidding until you notice that the left hand side is a perfect derivative.

Since $\dfrac{d}{dx}(\cos x) = -\sin x$, it follows (using the product rule) that

$$\frac{d}{dx}(y\cos x) = \frac{dy}{dx}\cos x - y\sin x. \qquad \left(\begin{array}{l}\text{This is the left hand side} \\ \text{of the differential equation.}\end{array}\right)$$

So we can rewrite the differential equation as

$$\frac{d}{dx}(y\cos x) = x^2.$$

We may now integrate both sides to obtain

$$y\cos x = \int x^2\,dx = \frac{x^3}{3} + c.$$

Dividing both sides by $\cos x$, the general solution is

$$y = \frac{x^3}{3\cos x} + \frac{c}{\cos x}.$$

In the previous example the left hand side was already a perfect derivative. That is not the case in the next example but it is a simple matter to convert it into one.

EXAMPLE

Find the general solution of the differential equation

$$\frac{dy}{dx} + \frac{2}{x}y = \frac{4}{x^2} \quad \text{for } x \neq 0.$$

Solution

First we note that the equation is linear, because it can be written in the form

$$\frac{dy}{dx} + Py = Q$$

where $P = \dfrac{2}{x}$ and $Q = \dfrac{4}{x^2}$. There are no terms in y^2, $\dfrac{1}{y}$, $\sqrt{y}$ etc.

If we now take the step of multiplying through by x^2, the equation becomes

$$x^2\frac{dy}{dx} + 2xy = 4.$$

The left hand side of this equation is now the expression you obtain when you differentiate x^2y with respect to x, using the product rule:

$$\frac{d}{dx}(x^2y) = x^2\frac{dy}{dx} + 2xy.$$

So the differential equation can be rewritten as

$$\frac{d}{dx}(x^2y) = 4.$$

Now integrating both sides gives

$$x^2y = \int 4\,dx = 4x + c.$$

The general solution is

$$y = \frac{4}{x} + \frac{c}{x^2}.$$

In each of these examples differential equations could be rewritten in the form

$$\frac{d}{dx}(Rx) = \text{function of } x$$

where R was some function of x. In the first example R = cos x, and in the second, R = x^2. Once the differential equation was written in this form, it was a straightforward task to solve it, since all that remained was to integrate the function of x on the right hand side.

However in the second example we had to multiply each term in the equation by a factor x^2 to bring the left hand side into the required form. This factor of x^2 is an example of an *integrating factor*; multiplying by it made the left hand side a perfect derivative.

Differential equations

Activity

Write the following differential equations in the form

$$\frac{d}{dx}(Rx) = \text{function of } x$$

and then find their general solutions.

(i) $x^3 \dfrac{dy}{dx} + 3x^2 y = 2e^x$ (ii) $e^{3x} \dfrac{dy}{dx} + 3e^{3x} y = x$

(iii) $\dfrac{dy}{dx} - \dfrac{1}{x} y = x^2.$

Part (iii) of the activity was harder than the others because the differential equation was written in the standard form

$$\frac{dy}{dx} + Py = Q.$$ *Notice that the coefficient of $\frac{dy}{dx}$ is 1.*

When the equation is in this form, you need to multiply each term by a function R of x, so that it becomes

$$R\frac{dy}{dx} + RPy = RQ$$ *Notice that the right hand side is still a function of x only.*

The function R has to be chosen such that the left hand side can be rewritten as $\dfrac{d}{dx}(Ry)$. You must therefore choose R such that

$$\frac{d}{dx}(Ry) \equiv R\frac{dy}{dx} + RPy.$$

Differentiating the left hand side, this becomes

$$R\frac{dy}{dx} + \frac{dR}{dx}y \equiv R\frac{dy}{dx} + RPy.$$

This is true if $\dfrac{dR}{dx} = RP,$

or $\dfrac{1}{R}\dfrac{dR}{dx} = P.$

This can be integrated by separating the variables to give

$$\ln R = \int P \, dx$$

$$\text{and so } R = e^{\int P \, dx}.$$

This means that any first order linear equation written in the standard form $\dfrac{dy}{dx} + Py = Q$ can be multiplied by an *integrating factor* $R = e^{\int P \, dx}$ to convert it to the compact form $\dfrac{d}{dx}(Ry) = RQ.$ You can then solve the equation by integrating the right hand side, which is a function of x only.

EXAMPLE

Solve the differential equation

$$x\frac{dy}{dx} + 2y = \frac{4}{x} \text{ for } x \neq 0.$$

Solution

Dividing through by x gives the equation in standard form:

$$\frac{dy}{dx} + \frac{2}{x}y = \frac{4}{x^2}.$$

> This is the equation that was solved by inspection on p. 40 but this time it is not presented in standard form.

Comparing this with $\frac{dy}{dx} + Py = Q$ gives $P = \frac{2}{x}$ and $Q = \frac{4}{x^2}$.

The integrating factor is $R = e^{\int P \, dx}$
$$= e^{\int \frac{2}{x} \, dx}$$
$$= e^{2 \ln x}$$
$$= e^{\ln x^2}$$
$$= x^2.$$

Multiplying the equation in standard form by this integrating factor gives

$$x^2 \frac{dy}{dx} + 2xy = 4.$$

The left hand side is now the derivative of a product, and can be written as $\frac{d}{dx}(Ry)$. In this case $R = x^2$ so that the left hand side is $\frac{d}{dx}(x^2y)$.

So the equation can be written

$$\frac{d}{dx}(x^2y) = 4$$

$$\Rightarrow \quad x^2y = 4x + c.$$

Dividing by x^2 gives the general solution

$$y = \frac{4}{x} + \frac{c}{x^2}.$$

This is of course the same solution as the one obtained on page 40. Now however, you have a method for finding an integrating factor, whereas previously you were relying on spotting a convenient form for the left hand side.

Now check that you can follow the stages in the next two examples.

EXAMPLE

Find the solution of the differential equation

$$x^2 \frac{dy}{dx} + xy = \frac{2}{x}$$

that satisfies the condition $y = 1$ when $x = 2$.

Solution

First the differential equation is written in standard form by dividing each term by x^2:

$$\frac{dy}{dx} + \frac{1}{x}y = \frac{2}{x^3}.$$

In this case $P = \frac{1}{x}$ and $Q = \frac{2}{x^3}$. The integrating factor is

$$R = e^{\int P\,dx} = e^{\int \frac{1}{x}\,dx} = e^{\ln x} = x.$$

Multiplying the differential equation by the integrating factor gives

$$x\frac{dy}{dx} + y = x \times \frac{2}{x^3}$$

or

$$\frac{d}{dx}(xy) = \frac{2}{x^2}.$$

Integrating with respect to x gives

$$xy = \int \frac{2}{x^2}\,dx$$

$$= -\frac{2}{x} + c$$

Dividing by x gives the general solution as

$$y = -\frac{2}{x^2} + \frac{c}{x}$$

The condition given is $y = 1$ when $x = 2$, and substituting these values in the general solution gives

$$1 = -\frac{2}{4} + \frac{c}{2}$$

and so $c = 3$.

The particular solution of the differential equation is

$$y = -\frac{2}{x^2} + \frac{3}{x}.$$

EXAMPLE

Use the integrating factor method to find the general solution of the differential equation

$$(\cos x)\frac{dy}{dx} - (\sin x)y = x^2$$

Solution

The equation is written in standard form by dividing through by $\cos x$:

$$\frac{dy}{dx} - \frac{\sin x}{\cos x}y = \frac{x^2}{\cos x}.$$

Here $P = -\frac{\sin x}{\cos x}$ and $Q = \frac{x^2}{\cos x}.$

The integrating factor is $R = e^{\left(\int -\frac{\sin x}{\cos x}\,dx\right)} = e^{(\ln(\cos x))} = \cos x.$

Multiplying through by cos x gives

$$\frac{dy}{dx}\cos x - y\sin x = x^2.$$

$$\Rightarrow \qquad \frac{d}{dx}(y\cos x) = x^2.$$

This is integrated to give

$$y\cos x = \frac{x^3}{3} + c.$$

Dividing through by cos x gives the general solution

$$y = \frac{x^3}{3\cos x} + \frac{c}{\cos x}.$$

This is the general solution of the differential equation.

NOTE

The integrating factor method involves several steps that can introduce errors on the way through, so it is important to verify that the function you think is the answer does indeed satisfy the original differential equation.

HISTORICAL NOTE

There is some uncertainty as to who discovered integrating factors. Some claim the credit for Leibniz who died in 1716 but the invention is generally believed to have been made somewhat later by Euler in 1734 and, independently, by Fontaine and Clairaut.

Claude Clairaut was a child prodigy who wrote a paper on third order curves at the age of 10 and followed it up shortly afterwards with one on the differential geometry of twisted curves in space; his brother was also a prodigy but died at the age of 16 having already completed original work on geometry.

Exercise 4A

1. Find the integrating factor for each of the following differential equations.

(i) $\frac{dy}{dx} + x^2 y = x$

(ii) $\frac{dy}{dx} + y\sin x = x$

(iii) $4x\frac{dy}{dx} - y = x^2$

(iv) $x^2\frac{dy}{dx} + xy = 2$

(v) $\frac{dy}{dx} + 7y = 1$

(vi) $\cos x \frac{dy}{dx} + y\sin x = e^{-2x}$

2. Find the particular solution of each of the following differential equations.

(i) $x\frac{dy}{dx} + 2y = x^2;$ $y = 0$ when $x = 1.$

(ii) $\frac{dy}{dx} + xy = 4x;$ $y = 2$ when $x = 0.$

(iii) $6xy + \frac{dy}{dx} = 0;$ $y = 3$ when $x = 1.$

(iv) $\frac{dy}{dx} - 2xy = x;$ $y = 1$ when $x = 0.$

(v) $\dfrac{dy}{dx} - \dfrac{3}{x}y = x;$ $y = 0$ when $x = 1.$

(vi) $\dfrac{dy}{dx} + 3x^2 y = x^2;$ $y = -1$ when $x = 0.$

3. An object falling vertically experiences air resistance so that the velocity satisfies the differential equation

$$\frac{dv}{dt} = 10 - 0.4v$$

(i)   Use the integrating factor method to find the general solution of this differential equation.
(ii) If initially $v = 0$, find the particular solution.
(iii) Solve the equation using the method of separation of variables.
(iv) Compare your two solutions.
(v)   If a linear differential equation is separable, which of the two methods would you prefer to use, the integrating factor method or the method of separation of variables?

4. A parachutist has a terminal speed of 30 ms^{-1}. The magnitude of the air resistance acting when the parachute is open is modelled by $F = kmv$ newtons, where k is a constant, v is the speed and m is the mass of the parachutist.
(i)   Find the value for k, taking the acceleration due to gravity to be 10 ms^{-2}.
(ii) Formulate a first order differential equation for the velocity, v.

(iii) Use the integrating factor method to find the general solution of the differential equation.
(iv) Find the particular solution if the parachutist is moving at 60 ms^{-1} when the parachute opens.

5. The radioactive isotope uranium 238 decays into thorium 234 which in turn decays into protactinium 234. This can be summarised as

$$\begin{array}{ccccc} 238 & & 234 & & 234 \\ \text{U} & \overset{k_1}{\rightarrow} & \text{Th} & \overset{k_2}{\rightarrow} & \text{Pa} \\ 92 & & 90 & & 91 \end{array}$$

where k_1 and k_2 are reaction constants ($k_1 \neq k_2$). The amounts of uranium 238, thorium 234 and protactinium 234 at time t are denoted by x, y and z. An experiment begins with an amount a of uranium 238 present, but no thorium 234 or proactinium 234. The amount y of thorium 234 present at time t satisfies the differential equation

$$\frac{dy}{dt} + k_2 y = k_1 a e^{-k_1 t}.$$

(i)   Find the integrating factor for the differential equation and hence its general solution.
(ii) Find the particular solution that satisfies the initial conditions.
(iii) Write down differential equations governing x and z.
(iv) (You may assume that the rate of decay of an isotope is proportional to the amount present).
The differential equation given in this problem is common in radioactive decay. Investigate what would happen in a case where $k_1 = k_2$.

4

KEY POINTS

- Any first order linear differential equation can be written in the form

$$\frac{dy}{dx} + Py = Q$$

where P and Q are functions of x only.

- To solve the equation $\frac{dy}{dx} + Py = Q$ you multiply throughout by the

integrating factor $R = e^{\int P \, dx}$. The solution is then given by

$$Ry = \int RQ \, dx.$$

5

Euler's Method

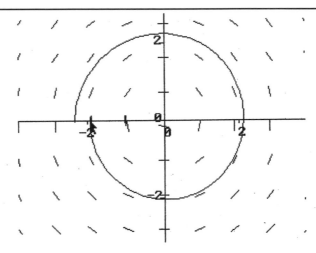

The picture shows a computer-generated tangent field for the differential equation

$$\frac{\mathrm{d}y}{\mathrm{d}x} = -\frac{x}{y},$$

together with the particular solution curve with initial condition $y = 0$ **when** $x = -2$.

(i) **Solve the equation analytically and so find the equation of the required particular solution. Describe the curve this equation represents.**

(ii) **How can you tell that the computer has gone wrong? Why has it happened?**

There are many first order differential equations which cannot be solved analytically. Look at the non-linear differential equation

$$\frac{\mathrm{d}y}{\mathrm{d}x} = x^2 + y^2.$$

You cannot separate the variables in this equation because it is not of the form

$$\frac{\mathrm{d}y}{\mathrm{d}x} = \mathrm{f}(x)\mathrm{g}(y).$$

Neither can you use an integrating factor, because the equation is not of the linear form

$$\frac{\mathrm{d}y}{\mathrm{d}x} + \mathrm{P}y = \mathrm{Q}.$$

Although there are many more analytical methods for solving first order differential equations than are covered in this book, it is often difficult or even impossible to solve a differential equation by analytical methods. In this case a numerical method of solution must be used.

In this chapter you will meet the simplest numerical method for solving differential equations. It is called *Euler's method*.

Straight line segments

The idea behind Euler's method is to approximate the solution curve (which you cannot find exactly) by a sequence of straight lines. The gradient of each straight line is the same as the gradient of the direction indicator in the tangent field at the start of the line.

As an example, take the differential equation

$$\frac{dy}{dx} = x^2 + y^2$$

with initial condition $y = 1$ when $x = 1$. Since you cannot solve the differential equation analytically, you cannot find an equation for the solution curve through the point $(1,1)$. Euler's method allows you to find an approximation to the actual curve, using the initial conditions and the information you have about the gradient of the curve.

The slope of the direction indicator at the point $(1,1)$ is given by

$$\frac{dy}{dx} = 1^2 + 1^2 = 2.$$

Figure 5.1 shows a line segment AB starting at point A $(1,1)$ with gradient 2. The broken line represents the actual (but unknown) solution curve through $(1,1)$. You can see that the line segment is a tangent to this solution curve at $(1,1)$. This line segment is the first part of your approximate curve. To find the next part of the curve you need a new starting point.

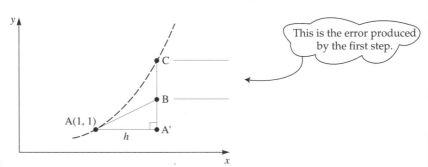

Figure 5.1

After a step size h in the x direction, the exact solution is an unknown point C and the estimate using the straight line segment is at B. The error in using B instead of C is BC.

If $h = 0.2$ then you can see that in triangle AA'B

$$\frac{A'B}{AA'} = \text{gradient of tangent } AB = 2$$

$$A'B = 2AA' = 2h = 0.4.$$

The estimates for the co-ordinates of point B are $x = 1.2$ and $y = 1.4$. The next line segment will start from B (1.2, 1.4) and have the same gradient as the direction indicator at B. It is shown as the line BD in figure 5.2. The gradient of BD is $1.2^2 + 1.4^2 = 3.4$.

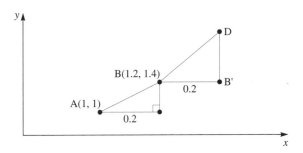

Figure 5.2

The co-ordinates of point D can be obtained from triangle BB'D in figure 5.2, using the fact that the gradient of BD is 3.4:

$$DB' = 0.2 \times 3.4 = 0.68.$$

D is therefore the point (1.4, 2.08).

Figure 5.3 shows such a process over several steps. Usually the further you move away from the starting point the larger is the error between the sequence of line segments and the actual solution curve. The error can usually be reduced by taking smaller steps.

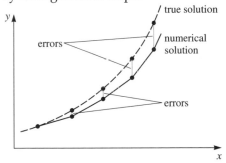

Figure 5.3

You will realise that drawing a sequence of straight lines from a tangent field is a rather painstaking and slow method. The advantage of Euler's method is that we can formulate a simple rule to generate the co-ordinates of points B, C, ... numerically as a table of values. This means that the process can easily be programmed into a calculator or computer, or carried out on a spreadsheet.

Suppose you have a tangent field defined by the differential equation

$$\frac{dy}{dx} = f(x, y)$$

and after n steps you have estimated the value of y at x_n to be y_n. With a step length of h, and with (x_0, y_0) as the initial condition for the problem, you can see that

$$x_n = x_0 + nh.$$

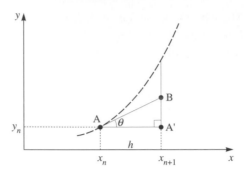

Figure 5.4

In Figure 5.4, point A has co-ordinates (x_n, y_n) and AB is the line segment whose gradient matches that of the direction field at A. The slope of AB is therefore $f(x_n, y_n)$. Point B has co-ordinates (x_{n+1}, y_{n+1}). The value of y_{n+1} is calculated as follows.

From triangle AA'B,

$$\tan\theta = \frac{A'B}{AA'} = \frac{y_{n+1} - y_n}{h}$$

$$\Rightarrow \quad y_{n+1} = y_n + h\tan\theta$$

But at A, $\qquad \tan\theta = \dfrac{dy}{dx} = f(x_n, y_n)$

$$\Rightarrow \quad y_{n+1} = y_n + hf(x_n, y_n).$$

This is *Euler's formula*. You can use it to calculate the y values of the solution to the differential equation once you know the initial values of x and y. It is an example of a *step-by-step method* because you proceed one small step at a time to find a solution. For each new step you use the results of the last step. The following example shows the method in use.

EXAMPLE

Use Euler's formula to estimate the value at $x = 1$ of the particular solution of the differential equation

> You saw how to construct a tangent field for this equation on p. 21.

$$\frac{dy}{dx} = x^2 + y^2$$

which passes through the point $(0, 0.5)$. Use a step length of 0.25.

Solution

It is a good idea to lay out the solution as a table, showing all the calculations that are performed. The following table shows the results rounded to 4 decimal places.

Differential equations

n	x_n	y_n	$f(x_n, y_n)$	$hf(x_n, y_n)$	$y_{n+1} = y_n + hf(x_n, y_n)$
0	0	0.5	0.25	0.0625	0.5625
1	0.25	0.5625	0.3789	0.0947	0.6572
2	0.5	0.6572	0.6819	0.1705	0.8277
3	0.75	0.8277	1.2476	0.3119	1.1396
4	1.0	1.1396			

This means the value of y when $x = 1.0$

Euler's numerical method gives the estimate $y(1.0) = 1.1396$.

The calculations involved in Euler's method (as for many other numerical methods) are easily carried out on a spreadsheet. Table 5.1 shows how the spreadsheet might look on the screen (for $h = 0.1$). One advantage of a spreadsheet approach is that you can easily change the step length and generate a new set of values. Consequently it is easy to see how the error is reduced by using a smaller step length.

Spreadsheet for Euler's method				$y' = x^2 + y^2$	
Please input the following variables:					
start for:					
x	0				
y	0.5				
step length (h)	0.1				
n	xn	yn	f(xn,yn)	hf(xn.yn)	yn+1=yn+hf(xn,yn)
0	0	0.5	0.25	0.025	0.525
1	0.1	0.525	0.285625	0.0285625	0.5535625
2	0.2	0.5535625	0.3464314	0.0346431	0.5882056
3	0.3	0.5882056	0.4359859	0.0435986	0.6318042
4	0.4	0.6318042	0.5591766	0.0559177	0.6877219
5	0.5	0.6877219	0.7229614	0.0722961	0.760018
6	0.6	0.760018	0.9376274	0.0937627	0.8537808
7	0.7	0.8537808	1.2189416	0.1218942	0.9756749
8	0.8	0.9756749	1.5919416	0.1591942	1.1348691
9	0.9	1.1348691	2.0979279	0.2097928	1.3446619
10	1	1.3446619	2.8081156	0.2808116	1.6254734

10 steps of size 0.1 give this answer.

Table 5.1: $f(x_n, y_n) = x_n^2 + y_n^2$; $h = 0.1$

Activity

Use a spreadsheet to find, correct to 2 decimal places, the value of y when $x = 1$ for $\dfrac{dy}{dx} = x^2 + y^2$, given that $y = 0.5$ when $x = 0$. The display for $h = 0.1$ should be like the one above, but you will need then to take smaller steps to achieve the required accuracy.

DE

Exercise 5A

Each question gives a differential equation, an initial condition and a step length. Use Euler's method with the given step length, h, to estimate the value of y for the stated value of x. Repeat each solution with smaller step lengths and compare your answers.

1. $\dfrac{dy}{dx} = x - y^2$; $\quad y = 1$ when $x = 2$.

Using $h = 0.2$, estimate the value of y (3) .

2. $\dfrac{dy}{dx} = \sqrt{x + y}$; $y = 2$ when $x = 1$.

Using $h = 0.1$, estimate the value of y (1.4).

3. $\dfrac{dy}{dx} = x^2 - y^2$; $\quad y = 2.1$ when $x = 1.5$.

Using $h = 0.2$, estimate the value of y (2.5).

4. $\dfrac{dy}{dx} = 2xy + 1$; $y = 0$ when $x = 0$.

Using $h = 0.1$, estimate the value of y (1).

Analysing the error

You probably sense that the smaller the step size, the better is the estimate you obtain using Euler's method. We can illustrate this by using Euler's method on a differential equation which has an analytical solution.

Take the simple differential equation $\dfrac{dy}{dx} = y$, with initial condition $y = 1$

when $x = 0$. The analytical solution is $y = e^x$.

The formula for Euler's method is

$$y_{n+1} = y_n + hf(x_n, y_n)$$

and in this case this becomes

$$y_{n+1} = y_n + hy_n \quad \text{with } y_0 = 1.$$

> Since $f(x, y) = y$ in this case, $f(x_n, y_n) = y_n$.

Table 5.2 shows the approximations to $y(1)$ obtained using different values of h. It also shows the percentage error calculated for each case using

$$\% \text{ error } = \frac{\text{analytical solution } - \text{ numerical solution}}{\text{analytical solution}} \times 100.$$

In this case the analytical solution is known to be $y(1) = e^1 = 2.718282$ (to 6 decimal places).

h	numerical solution	% error	Number of steps
0.1	2.593742	4.58	10
0.01	2.704814	0.50	100
0.001	2.716924	0.05	1000
0.0001	2.718146	0.005	10000

Table 5.2

The table shows that the accuracy is much increased if you reduce the step size from 0.1 to 0.0001. However, this clearly lengthens the time needed for the calculation: it takes 10 steps of calculation if $h = 0.1$ and 1000 steps if $h = 0.001$. To obtain a sufficiently accurate estimate may require a very small step size, necessitating additional calculations and more computer time.

Of course, the importance of a numerical method is not its use in solving differential equations that can be solved analytically, but in solving differential equations for which analytical methods do not exist. Table 5.3 shows the estimate of $y(1.0)$ obtained by applying Euler's numerical method to the differential equation $\dfrac{dy}{dx} = x^2 + y^2$ (which cannot be solved analytically), with the initial condition $y = 0.5$ when $x = 0$.

h	Estimate of y (1.0)
0.2	1.1964
0.125	1.3020
0.1	1.3447
0.0625	1.4185
0.05	1.4463
0.025	1.5081
0.01	1.5498

Table 5.3

It is not possible to do the calculation with a step length of zero, because this would involve an infinite number of calculations. However, were you able to do so, you would expect the answer to be exact, with zero error (ignoring any computer-generated error). You could express this as 'when $h \rightarrow 0$, error $\rightarrow 0$'.

A graph of the estimate of $y(1.0)$ against h is shown in figure 5.5.

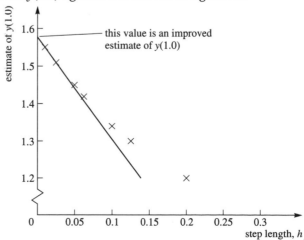

Figure 5.5 Graph of estimates of $y(1.0)$ for different step lengths

DE

Figure 5.5 illustrates an important feature of Euler's method. As the step size decreases, the estimates for $y(1.0)$ appear to lie very close to a straight line.
• For small values of h the error is approximately a linear function of h.

This provides a useful method of increasing the accuracy for the estimate of y using Euler's method, with little increase in the number of computations or the amount of computer time.

For discussion

Look again at the graph in figure 5.5. The straight line that has been drawn is deliberately not a line of best fit. Why not?

EXAMPLE

You are given the differential equation

$$\frac{dy}{dx} = \frac{y}{x} + 2$$

with initial condition $y = 1$ when $x = 1$.
(i)   Estimate the value of $y(2)$ using Euler's method with step size $h = 0.1$ and then with step size $h = 0.05$.
(ii) Use the linear relationship between the error and h in Euler's method to obtain an improved estimate for the value of $y(2)$.
(iii) The analytical solution of the differential equation is

$$y = x + 2x \ln x.$$

Compare your estimate of $y(2)$ in part (ii) with the exact value. Estimate the step size that is required for the basic Euler's method to give an answer correct to four decimal places.

Solution
(i)   The formula for Euler's method in this case is

$$y_{n+1} = y_n + h\left(\frac{y_n}{x_n} + 2\right)$$

where $x_n = 1 + nh$, with initial conditions $y_0 = 1$ and $x_0 = 1$.
The estimates of $y(2.0)$ obtained using $h = 0.1$ and $h = 0.05$ are shown below.

h	Estimate for $y(2.0)$
0.1	4.67508
0.05	4.72321

(ii) If there is a linear relationship between $y(2)$ and h, then it can be expressed using the straight line equation $y(2) = ah + b$, where a and b are constants. Substituting into this equation the solutions for $h = 0.1$ and $h = 0.05$, you can find values for a and b:

$$4.67508 = 0.1a + b$$
$$4.72321 = 0.05a + b$$

Subtracting the second equation from the first,

$$-0.04813 = 0.05a,$$

So $a = -0.96260$ and $b = 4.77133$.

This gives the equation relating $y(2)$ and h as

$$y(2) = 4.77133 - 0.9626h.$$

Using this equation you can estimate the exact value of $y(2)$ by putting $h = 0$, giving

$$y(2) = 4.77133.$$

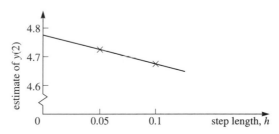

(iii) The analytical solution gives $y(2) = 2 + 4 \ln 2 = 4.77258$.

The numerical solution in part (ii) is correct to two decimal places. The error in Euler's method is proportional to the step size h, so the error can be expressed as a function of h:

$$y_E - y_N = ch$$

where y_E is the exact solution, y_N is the numerical solution for step size h and c is a constant. Using the solution for $h = 0.05$ and the analytical solution 4.77258, you can find a value for c:

$$4.77258 - 4.72321 = c \times 0.05$$
$$\Rightarrow \qquad c = 0.987398.$$

To obtain an accuracy of four decimal places, you require

$$y_E - y_N \leq 0.00005$$
$$\Rightarrow \qquad ch \leq 0.00005$$
$$\Rightarrow \qquad h \leq \frac{0.00005}{0.987398}$$
$$\textit{i.e.} \qquad h \leq 5.1 \times 10^{-5}.$$

This small step size shows that almost 20 000 steps are needed to achieve an accuracy of four decimal places.

NOTE

In the circumstances in which you would use a numerical method you would not know the exact answer. Instead you would estimate your accuracy by looking at the sequence of estimates you had obtained with progressively smaller step lengths. This means that at any stage in the process, the level of accuracy that you have actually achieved is higher than you are able to justify.

In this chapter you have seen Euler's method in action, and seen that the error in the approximate solution can be decreased by reducing the step length. You have also seen that the error is roughly proportional to the step length, providing the step length is small. However, this is only part of the story. It is justifiable to take the error to be proportional to the step length h only when h is small. However, working with small values of h can introduce another source of error if your calculator or computer is working to only a limited number of figures. The area of mathematics that deals with these issues in detail is called *numerical analysis*.

Leonhard Euler was one of the most prolific and versatile mathematicians ever, his life-time's work amounting to 886 books and papers. Euler was born in Basel in Switzerland in 1707 and as a boy studied under Johann Bernoulli. In 1727 he took the chair of mathematics at St Petersburg academy in Russia; he later moved to the Prussian Academy in Berlin but returned to St Petersburg for his closing years. He died in 1783 at the age of 76.

Euler's contributions to mathematics are wide-ranging, including the introduction of the symbols $f(x)$, e, Σ and the complex numbers results $e^{jx} = \cos x + j\sin x$ and $e^{j\pi} + 1 = 0$. Much of his work was in applied mathematics including celestial mechanics, hydraulics, ship-building, artillery and music.

Exercise 5B

1. This problem is about the differential equation

$$\frac{dy}{dx} = 4x - 0.05y,$$

with the initial condition $y = 80$ when $x = 5$.

(i) Use Euler's numerical method to estimate the value of $y(6)$, using step lengths of 1.0, 0.5, 0.2 and 0.1.

(ii) Using your estimates in part (i) draw a graph, and from your graph obtain an improved estimate of the value of $y(6)$ as $h \to 0$.

(iii) Use the integrating factor method to find an analytical solution of the differential equation, and hence an exact value of $y(6)$. Compare this with your answer to part (ii).

(iv) Calculate the step size and number of steps needed to obtain an estimate correct to 3 decimal places.

2. This problem is about the differential equation

$$\frac{dy}{dx} = xy^2,$$

with initial condition $y = 0.5$ when $x = 1$.

(i) Use Euler's numerical method to estimate the value of $y(2)$, using step lengths of 0.2, 0.1, 0.05 and 0.01.

(ii) Use your estimates in part (i) to draw a graph, and from your graph obtain an improved estimate of the value of $y(6)$ as $h \to 0$.

(iii) Use the method of separation of variables to find an analytical solution and hence an exact value of $y(2)$. Compare this with your answer to part (ii).

(iv) Calculate the step size and number of steps needed to obtain an estimate correct to 4 decimal places.

3. Estimate to 4 decimal places the value of $y(1)$ for

$$\frac{dy}{dx} = y + e^{-x} \sin x,$$

with initial condition $y = 1$ when $x = 0$.

4. Estimate to 4 decimal places the value of $y(1.5)$ for

$$\frac{dy}{dx} = 1 + x \sin y$$

with initial condition $y = 0.5$ when $x = 0$.

5. (i) Use Euler's method with step length 0.1 to estimate the value of $y(1.2)$ given the differential equation

$$\frac{dy}{dx} = x^2 - y^2 \qquad (x \geq 0)$$

with initial condition $y = 0$ when $x = 1$.

(ii) The table below shows approximate values of $y(1.2)$ for various step lengths.

Step length h	$y(1.2)$
0.01	0.237287
0.005	0.238220
0.002	0.238778
0.001	0.238964

Draw a graph of these approximations to $y(1.2)$ against step length h. What does your graph suggest to you about the relationship between the Euler approximation to $y(1.2)$ and the step length h for small h? Use the graph to estimate the value of $y(1.2)$. How many decimal places can you justify?

6. *In this question take g to be 10 ms^{-2}.*
When an object of mass m kg falls vertically under the influence of gravity it experiences two forces, its weight mg and a force F N due to air resistance. Its downward velocity v at time t s is thus governed by the differential equation

$$\frac{dv}{dt} = g - \frac{1}{m} F.$$

A particular object of mass 0.05 kg is found to have terminal velocity 20 ms^{-1} but the relationship between F and v is unclear.

Three models are considered:

(a) $F = k_1 v$ (b) $F = k_2 v^2$
(c) $F = k_3 v^{1.5}$

(i) Calculate the values of k_1, k_2 and k_3 corresponding to these three models.

In an experiment the object is dropped from rest from a great height.

(ii) Solve the differential equation corresponding to model (a) and draw the graph of v against t for values of t up to 12.

(iii) Verify that

$$v = 20 \frac{(1 - e^{-t})}{(1 + e^{-t})}$$

is a solution for model (b) and draw the graph of v against t on the same axes.

(iv) Use Euler's method to find suitable points to plot on the graph to illustrate model (c), $F = k_3 v^{1.5}$.

(v) A further (and more realistic) model is considered in which $F = k_4 v + k_5 v^2$. Explain why you are unable to illustrate this model on the graph as well as the others.

Exercise 5B continued

7. A car headlight mirror is designed to reflect the light from the headlamp in rays which are parallel to the road surface. In this problem you investigate the required cross-section shape of the headlight mirror. The first diagram shows a geometric model of the situation in which the light rays are emitted from a point source, at the origin.

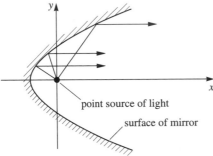

In the second diagram, P is a general point (X, Y) on the curve. The line PQ is the tangent at P, with Q on the x axis. The angle OPQ = θ. The point T is positioned as shown on the x axis. If the mirror is designed correctly, light travelling from the point source at O along OP will be reflected along a horizontal line from P, as shown.

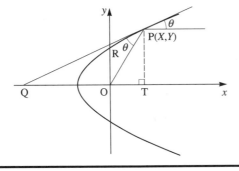

The laws of optics require that the angle of incidence is equal to the angle of reflection. Angle OPQ is the angle of incidence, marked as θ. The angle of reflection has also been marked as θ in the diagram.

(i) Write down the size of angle PQO and hence show that the length OQ is

$$\sqrt{\left(X^2 + Y^2\right)}.$$

(ii) Using triangle QPT, deduce that

$$\tan\theta = \frac{Y}{\sqrt{(X^2 + Y^2)} + X}.$$

(iii) Hence show that the gradient of the mirror at any point is given by

$$\frac{dy}{dx} = \frac{y}{\sqrt{(x^2 + y^2)} + x}.$$

The mirror is designed so that the distance OR of the mirror above the lamp is 6 cm. Taking 1 cm to be 1 unit, this can be written as $y(0) = 6$.

(iv) Use Euler's method to estimate the co-ordinates of several points along the curve.

(v) Use your answers to part (iv) to sketch the curve.

(vi) Verify that $y^2 = 12(x + 3)$ is the analytical solution of the differential equation. Describe the shape of the mirror.

Investigations

The Modified Euler Method

This method for solving the first order differential equation

$$\frac{dy}{dx} = f(x, y)$$

uses the formula $\quad y_{n+1} = y_n + \frac{1}{2}(k_1 + k_2)$

where $k_1 = hf(x_n, y_n)$ and $k_2 = hf(x_n + h, y_n + k_1)$.

The error in Euler's method is linearly dependent on h. Investigate, by solving several differential equations, each with a range of values of h, how the error in using the Modified Euler Method varies with h.

KEY POINTS

- Euler's numerical step-by-step method for solving the differential equation $\frac{dy}{dx} = f(x, y)$ with initial condition $y = b$ when $x = a$ and a step length h is given by

$$y_{n+1} = y_n + hf(x_n, y_n); \quad y_0 = b$$
$$x_n = a + nh$$

- The error in the estimate obtained from the numerical solution is approximately proportional to the step size h for small h.
- Reducing the step size usually reduces the error.

6 Linear equations with constant coefficients

As is well-known, Physics became a science only after the invention of differential calculus.

Riemann

How long does it take a parachutist to reach the ground?

A reasonable model is provided by considering the parachutist as a particle of mass m subject to two forces, weight, mg, and resistance, kv (where v is the speed and k a constant). Applying Newton's Second Law with the downward direction as positive gives

$$mg - kv = ma. \qquad ①$$

If the distance fallen is denoted by s, then

$$v = \frac{\mathrm{d}s}{\mathrm{d}t} \qquad \text{and} \qquad a = \frac{\mathrm{d}v}{\mathrm{d}t} = \frac{\mathrm{d}^2 s}{\mathrm{d}t^2}.$$

positive direction

So equation ① can be rewritten as

$$mg - k\frac{\mathrm{d}s}{\mathrm{d}t} = m\frac{\mathrm{d}^2 s}{\mathrm{d}t^2},$$

or

$$\frac{\mathrm{d}^2 s}{\mathrm{d}t^2} + \frac{k}{m}\frac{\mathrm{d}s}{\mathrm{d}t} = g. \qquad ②$$

Since this contains the term $\dfrac{\mathrm{d}^2 s}{\mathrm{d}t^2}$, it is a *second order* differential equation.

Higher order equations

So far in this book you have solved only first order equations. In this chapter we start to turn our attention to second (and higher) order equations. Second order equations are often needed in mechanics, particularly to model situations which (like the one above) involve acceleration.

Activity

Show that the differential equation ② for the parachutist opposite can be written as

$$\frac{d^2s}{dt^2} + 2\frac{ds}{dt} = 10$$

in the case where, in SI units, $m = 60$, $k = 120$ and $g = 10$.

Given that when $t = 0$, $s = 0$ and $\dfrac{ds}{dt} = 20$, verify that

$$s = 5t + 7.5(1 - e^{-2t})$$

is the particular solution of this equation.

Classification of differential equations

The differential equation

$$\frac{d^3y}{dx^3} - 7\frac{d^2y}{dx^2} + 2\frac{dy}{dx} + 4y = 3\sin x$$

is described as

- *third order* (because the highest derivative is third order);
- *linear* (because where y appears it is to the power 1: there are no terms involving y^2, $\sqrt{y}$ etc.);
- having *constant coefficients* (because the coefficients of the terms involving y, namely 1, –7, 2 and 4 are all constants);
- *non-homogeneous* (because the right hand side, the part not containing y, is not zero. In cases when it is zero, the equation ① is called *homogeneous*).

The general second order linear differential equation with constant coefficients may be written as

$$\frac{d^2y}{dx^2} + a\frac{dy}{dx} + by = f(x)$$

where a and b are constant. Although this equation appears quite daunting, it is in fact possible to solve it using a method which does not involve any integration. You will find that such equations arise quite often when you are modelling real situations. For this reason, much of the remainder of this book concentrates on solving second order equations of this type: Chapters 7 and 8 are devoted to a very important context in which they arise – that of oscillating systems.

NOTE

In many other textbooks the terms homogeneous *and* non-homogeneous *are taken to have a different meaning. The sense in which they are used in this book is now in widespread use.*

The first order equation

Before solving second and higher order differential equations you will find it helpful to look at the form of the solution of first order linear, homogeneous differential equations with constant coefficients. The general first order equation of this type can be written as

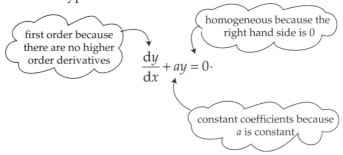

first order because there are no higher order derivatives

homogeneous because the right hand side is 0

$$\frac{dy}{dx} + ay = 0.$$

constant coefficients because a is constant

You can solve this general differential equation using the method of separation of variables:

$$\frac{dy}{dx} = -ay$$

$$\Rightarrow \quad \int \frac{1}{y} \, dy = \int -a \, dx$$

$$\Rightarrow \quad \ln |y| = -ax + c \quad (c \text{ is a constant of integration}).$$

This gives

$$y = \pm \, e^{(-ax + c)}$$
$$= A e^{-ax}$$

where $A = \pm \, e^c$ is a constant.

This shows that the solution of any first order equation of this type will contain an exponential function and an unknown constant.

Activity

Use the method of separation of variables to find the general solution of each of the following differential equations.

(a) $\dfrac{dy}{dx} + 2y = 0$ (b) $2\dfrac{dy}{dx} - 5y = 0$ (c) $3\dfrac{dx}{dt} + x = 0$

The auxiliary equation method

Knowing the form of the solution allows you to find it without doing any integration. This is demonstrated in the next example.

EXAMPLE

Solve the differential equation

$$5\frac{dy}{dx} + y = 0$$

Solution

We begin by assuming an exponential solution of the form

$$y = Ae^{\lambda x},$$

where λ and A are constants.

If this is to be a solution, it must satisfy the differential equation.

Differentiating y with respect to x,

$$\frac{dy}{dx} = \lambda Ae^{\lambda x}.$$

Substituting this into the differential equation gives

$$5\lambda Ae^{\lambda x} + Ae^{\lambda x} = 0.$$

Dividing both sides by $Ae^{\lambda x}$ gives an equation for λ:

$$5\lambda + 1 = 0. \qquad \qquad ①$$

Solving for λ,

$$\lambda = -\frac{1}{5}.$$

So the solution of the differential equation is

$$y = Ae^{-\frac{1}{5}x}.$$

The equation ① is called the *auxiliary equation* and this method is called the *auxiliary equation method*.

For discussion

This is in fact the *general* solution of this differential equation. How can you be sure that this is the case and there is not another possible solution?

Exercise 6A

1. Use the auxiliary equation method to find the general solutions of the following differential equations.

(i) $\dfrac{dy}{dx} - 3y = 0$ (ii) $\dfrac{dy}{dx} + 7y = 0$

(iii) $\dfrac{dx}{dt} + x = 0$ (iv) $\dfrac{dp}{dt} - 0.02p = 0$

(v) $5\dfrac{dz}{dt} - z = 0$

2. Use the auxiliary equation method to find the particular solutions of the following differential equations.

(i) $\dfrac{dy}{dx} + 2y = 0$; $y = 3$ when $x = 0$.

(ii) $2\dfrac{dy}{dx} - 5y = 0$; $y = 1$ when $x = 0$.

(iii) $3\dfrac{dx}{dt} + x = 0$; $x = 2$ when $t = 1$.

(iv) $\dfrac{dP}{dt} = kP$; $P = P_0$ when $t = 0$.

(v) $\dfrac{dm}{dt} = -km$; $m = m_0$ when $t = 0$.

3. Find the general solution of the differential equation

$$\frac{dy}{dx} + 5y = 0$$

using

(i) the method of separation of variables;

(ii) the integrating factor method;

(iii) the auxiliary equation method.

Show that in each case the answer can be written in the same form.

4. All of the following differential equations can be solved by at least one of the methods:

- separation of variables;
- integrating factor;
- auxiliary equation.

In each case state which method (or methods) can be used and use it (or them) to solve the equation.

(i) $\dfrac{dy}{dx} - 17y = 0$ (ii) $\dfrac{dy}{dx} - y^2 = 0$

(iii) $\dfrac{dy}{dx} + xy = 0$ (iv) $\dfrac{dy}{dx} - 3y = 0$

(v) $y\dfrac{dy}{dx} - y^2 = 0$

5. In each case, state what can be said about the function h(x, y) when the differential equation

$$\frac{dy}{dx} = h(x, y)$$

can be solved using:

(i)   the method of separation of variables;

(ii)   the integrating factor method;

(iii)   the auxiliary equation method.

Second order equations

On p. 63 we developed a method for solving first order equations based on our knowledge that their solutions have exponential form. Can this method be extended to second order equations?

Look at the equation

$$\frac{d^2y}{dx^2} - 5\frac{dy}{dx} + 6y = 0.$$

Suppose that we assume a solution of the form $y = Ae^{\lambda x}$ where A and λ are constants, then

$$\frac{dy}{dx} = A\lambda e^{\lambda x}$$

and

$$\frac{d^2y}{dx^2} = A\lambda^2 e^{\lambda x}.$$

Substituting these into the differential equation gives

$$A\lambda^2 e^{\lambda x} - 5A\lambda e^{\lambda x} + 6Ae^{\lambda x} = 0.$$

Dividing through by $Ae^{\lambda x}$, we obtain the quadratic equation in λ,

$$\lambda^2 - 5\lambda + 6 = 0,$$

and this is the *auxiliary equation*. Notice that the form of the auxiliary equation is very close to that of the original differential equation. It could have been written down straight away without any of the intermediate working.

Factorising the auxiliary equation gives
$$(\lambda - 3)(\lambda - 2) = 0.$$

You can see that there are two different values for λ which satisfy the auxiliary equation: $\lambda = 3$ and $\lambda = 2$. So $y = Ae^{3x}$ and $y = Be^{2x}$ (where A and B are constants) are two solutions of the differential equation

$$\frac{d^2y}{dx^2} - 5\frac{dy}{dx} + 6y = 0.$$

The expressions Ae^{3x} and Be^{2x} are called the *complementary functions* of the differential equation. The sum of these two expressions, $Ae^{3x} + Be^{2x}$, is usually called simply the *complementary function*.

You can now verify that this sum,

$$y = Ae^{3x} + Be^{2x}$$

is also a solution of the second order differential equation.

Differentiating this expression for y gives

$$\frac{dy}{dx} = 3Ae^{3x} + 2Be^{2x}$$

and

$$\frac{d^2y}{dx^2} = 9Ae^{3x} + 4Be^{2x}.$$

Substituting for $\dfrac{d^2y}{dx^2}, \dfrac{dy}{dx}$ and y, the left hand side of the original differential equation can now be written as

$$(9Ae^{3x} + 4Be^{2x}) - 5(3Ae^{3x} + 2Be^{2x}) + 6(Ae^{3x} + Be^{2x}).$$

After multiplying out the brackets this becomes

$$9Ae^{3x} + 4Be^{2x} - 15Ae^{3x} - 10Be^{2x} + 6Ae^{3x} + 6Be^{2x},$$

and you can see by collecting like terms that this is equal to zero for all x. So $y = Ae^{3x} + Be^{2x}$ is indeed a solution and this is in fact the general solution.

The method used in this example is one that can be used on any linear equation with constant coefficients, whatever its order. If $\lambda_1, \lambda_2, \lambda_3, \dots$ are the roots of the auxiliary equation, then the general solution of the homogeneous differential equation is

$$y = Ae^{\lambda_1 x} + Be^{\lambda_2 x} + Ce^{\lambda_3 x} + \dots$$

This is an example of the *principle of superposition*. If $u(x), v(x), w(x), \dots$ are solutions of a linear differential equation, then the general solution is

$$y = Au + Bv + Cw + \dots$$

where $A, B, C, \dots$ are arbitrary constants. The function y is said to be a *linear combination* of the functions $u, v, w, \dots$

The principle of superposition applies to any homogeneous differential equation with constant coefficients. The following example illustrates the method for a fourth order differential equation.

EXAMPLE

Solve the differential equation

$$\frac{d^4y}{dx^4} - 2\frac{d^3y}{dx^3} - \frac{d^2y}{dx^2} + 2\frac{dy}{dx} = 0$$

Solution

We can go straight to the auxiliary equation:

$$\lambda^4 - 2\lambda^3 - \lambda^2 + 2\lambda = 0$$

Factorising this quartic equation gives

$$\lambda(\lambda - 1)(\lambda + 1)(\lambda - 2) = 0$$

This means that there are four possible roots for λ: $\lambda = 0$, $\lambda = -1$, $\lambda = 1$, $\lambda = 2$.

The general solution is

$$y = A + Be^{-x} + Ce^{x} + De^{2x}$$

in which A, B, C and D are constants.

There are three important points to notice from this example:

- the method does not involve any integration;
- the number of complementary functions is equal to the order of the differential equation;
- the number of unknown constants in the solution is equal to the order of the differential equation.

Find the general solution of each of the
following differential equations.

1. $\dfrac{d^2y}{dx^2} - 3\dfrac{dy}{dx} + 2y = 0$

2. $6\dfrac{d^2y}{dx^2} + 5\dfrac{dy}{dx} + y = 0$

3. $2\dfrac{d^2y}{dx^2} + \dfrac{dy}{dx} - y = 0$

4. $\dfrac{d^2y}{dx^2} + 4\dfrac{dy}{dx} - 5y = 0$

5. $\dfrac{d^2y}{dx^2} - 4y = 0$

6. $\dfrac{d^3y}{dx^3} - 6\dfrac{d^2y}{dx^2} + 11\dfrac{dy}{dx} - 6y = 0$

7. $\dfrac{d^2x}{dt^2} + \dfrac{dx}{dt} - 6x = 0$

8. $\dfrac{d^2x}{dt^2} - 9x = 0$

9. $\dfrac{d^2u}{dt^2} - 3\dfrac{du}{dt} = 0$

10. $m\dfrac{d^2x}{dt^2} - kx = 0$

where m and k are positive constants.

Finding the values of the unknown constants

The general solution of a second order differential equation contains two
unknown constants. To find the values of these constants you need two extra
pieces of information. (Remember that for first order equations you needed
one extra piece of information.) The extra information is given in one of two
different ways.

Initial conditions

If the two pieces of information are given for the same value of the
independent variable, we say that we have two *initial conditions*.

For example, $y = 2$ at $x = 1$ and $\dfrac{dy}{dx} = 3$ at $x = 1$ are initial conditions
(at $x = 1$).

Problems consisting of a differential equation and initial conditions are called
initial value problems.

Boundary conditions

If the information is given at two values of the independent variable, we say
that we have two *boundary conditions*.

For example, $y = 2$ at $x = 1$ and $y = 1$ at $x = 3$ are boundary conditions
(at $x = 1$ and $x = 3$). Problems consisting of a differential equation and two
boundary conditions are called *boundary value problems*. In applications which
involve boundary value problems the solution is often restricted to the region
between the boundary points.

The following examples show how to use initial and boundary conditions to
find the unknown constants.

Linear equations with constant coefficients

EXAMPLE

(i) Find the particular solution of

$$\frac{d^2y}{dx^2} + 4\frac{dy}{dx} + 3y = 0$$

given the initial conditions $y = 0$ and $\frac{dy}{dx} = 1$ at $x = 0$.

(ii) Sketch the graph of the solution.

Solution

(i) The auxiliary equation is $\lambda^2 + 4\lambda + 3 = 0$, which has roots $\lambda = -3$ and $\lambda = -1$. The general solution of the differential equation is

$$y = Ae^{-x} + Be^{-3x}.$$

The first initial condition says that when $x = 0$, $y = 0$, so substituting these into the general solution gives

$$A + B = 0. \qquad ①$$

Differentiating the general solution gives

$$\frac{dy}{dx} = -Ae^{-x} - 3Be^{-3x}.$$

The second initial condition requires that when $x = 0$, $\frac{dy}{dx} = 1$.

Substituting these values into the expression for $\frac{dy}{dx}$ gives

$$-A - 3B = 1. \qquad ②$$

Solving equations ① and ② gives $A = \frac{1}{2}$ and $B = -\frac{1}{2}$.

The particular solution is $y = \frac{1}{2}e^{-x} - \frac{1}{2}e^{-3x}$.

(ii) The graph of y is shown in the diagram. Both of the terms

$\frac{1}{2}e^{-x}$ and $-\frac{1}{2}e^{-3x}$ decay to zero with increasing x.

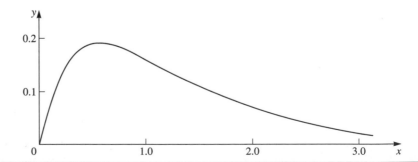

EXAMPLE

The differential equation

$$\frac{d^2y}{dx^2} + 4\frac{dy}{dx} + 3y = 0.$$

Notice that this is the same differential equation as in the last example.

is used to model a situation in which the value of x lies between 0 and 2.

(i) Find the particular solution given the boundary conditions $y = 0$ when $x = 0$ and $y = 1$ when $x = 2$.

(ii) Sketch a graph of the solution for $0 \leq x \leq 2$.

Solution

The general solution is

$$y = Ae^{-x} + Be^{-3x},$$

as before.

Using the boundary conditions:

when $x = 0$, $y = 0 \Rightarrow A + B = 0$ ①

when $x = 2$, $y = 1 \Rightarrow Ae^{-2} + Be^{-6} = 1$.   ②

Substituting for B from ① into ② gives

$$Ae^{-2} - Ae^{-6} = 1$$

$$A = \frac{e^2}{1 - e^{-4}}$$

and so

$$B = -\frac{e^2}{1 - e^{-4}}.$$

The particular solution in this case is

$$y = \frac{e^2}{1 - e^{-4}}(e^{-x} - e^{-3x})$$

The diagram shows the values of y for the given range of values of x.

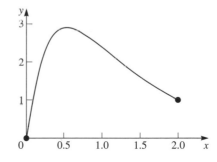

Exercise 6C

Find the particular solution of each of the differential equations in questions 1–5, and illustrate it with a sketch graph. Use a graphics calculator or a computer to check your answers. Describe the shape of the graph as the independent variable increases.

1. $\dfrac{d^2y}{dx^2} - 5\dfrac{dy}{dx} + 6y = 0;$

initial conditions $y = 1$ and

$\dfrac{dy}{dx} = 0$ when $x = 0$.

2. $\dfrac{d^2y}{dx^2} - 9y = 0;$

initial conditions $y = 0$ and

$\dfrac{dy}{dx} = 1$ when $x = 0$.

3. $\dfrac{d^2x}{dt^2} + 6\dfrac{dx}{dt} + 5x = 0;$

boundary conditions $x = 0$ when $t = 0$ and $x = 2$ when $t = 1$.

4. $\dfrac{d^2v}{dt^2} - v = 0;$

boundary conditions $v = -1$
when $t = -1$ and $v = 1$ when
$t = 1.$

5. $\dfrac{d^2y}{dx^2} - 5\dfrac{dy}{dx} = 0;$

initial conditions $y = 0$ and

$\dfrac{dy}{dx} = 4$ when $x = 0.$

6. The metal fins on a motor cycle engine
help to cool it. A model for the change of
temperature along a fin is given by

$$\dfrac{d^2T}{dx^2} - 4T = 0$$

where $T\ °C$ is the temperature and
x m is the distance from the hot end.

(i) Find the general solution of this
differential equation.

When the engine has been running for
some time the temperatures at the two
ends of the fin are $100\ °C$ and $80\ °C$. The
fin is 5 cm long.

(ii) Find the particular solution.

(iii) Hence determine the
temperature 3 cm from the hot end.

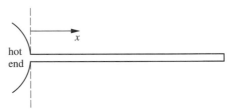

Auxiliary equation with complex roots: pure imaginary roots

The auxiliary equation for a second order differential equation is a quadratic
equation in λ. So far you have only seen cases in which the roots of the
auxiliary equation were real and distinct numbers λ_1 and λ_2. The general
solution of these differential equations had the form

$$y = Ae^{\lambda_1 x} + Be^{\lambda_2 x} .$$

However, the auxiliary equation does not always have real roots. For
example, the differential equation

$$\dfrac{d^2y}{dx^2} + 4y = 0 \qquad\qquad ①$$

has auxiliary equation

$$\lambda^2 + 4 = 0.$$

You can see that this has no real roots. The roots of the equation are the
complex numbers $\lambda_1 = 2j$ and $\lambda_2 = -2j$ (where $j = \sqrt{-1}$).
NOTE *You will see* i *used instead of* j *for* $\sqrt{-1}$ *in some texts.*

The general solution of differential equation ① is therefore

$$y = Ae^{2jx} + Be^{-2jx}.$$

Complex exponentials can be written in terms of cosine and sine functions using the relationships

$$e^{j\theta} = \cos\theta + j\sin\theta$$
$$e^{-j\theta} = \cos\theta - j\sin\theta$$

(An explanation of this is given in *Pure Mathematics 5* in this series.)

The general solution can therefore be written as

$$y = A(\cos 2x + j\sin 2x) + B(\cos 2x - j\sin 2x)$$
$$= (A + B)\cos 2x + (jA - jB)\sin 2x.$$

Since A and B are simply unknown constants, you can define new constants C and D such that $A + B = C$ and $jA - jB = D$. The general solution of the original differential equation

$$\frac{d^2y}{dx^2} + 4y = 0$$

can then be written simply as

$$y = C\cos 2x + D\sin 2x.$$

From now on we write this as

$$y = A\sin 2x + B\cos 2x$$

(which simply means that we have renamed the constants).

You can verify that this is the general solution by differentiating y and substituting the resulting expressions into the differential equation.

$$\frac{dy}{dx} = 2A\cos 2x - 2B\sin 2x$$

$$\frac{d^2y}{dx^2} = -4A\sin 2x - 4B\cos 2x.$$

Substitution into the left hand side of the differential equation gives

$$(-4A\sin 2x - 4B\cos 2x) + 4(A\sin 2x + B\cos 2x)$$

By collecting like terms you will see that this is zero for all values of x, so $y = A\sin 2x + B\cos 2x$ satisfies the differential equation.

Activities

1. Show that substituting $j\theta$ for x in the expansion

$$e^x = 1 + x + \frac{x^2}{2!} + \frac{x^3}{3!} + \dots + \frac{x^r}{r!} \dots$$

gives $\quad e^{j\theta} = \cos\theta + j\sin\theta.$

2. Show (i) by verification, and (ii) using the auxiliary equation method, that the general solution of the differential equation

$$\frac{d^2x}{dt^2} + \omega^2 x = 0$$

is $\quad x = A\sin\omega t + B\cos\omega t.$

Oscillating solutions

In cases where the auxiliary equation has real roots, the solution involves either exponential growth or exponential decay, as shown in figure 6.1.

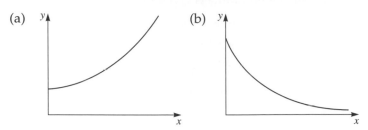

Figure 6.1 Typical solution curves for $y = Ae^{\lambda x}$ (where λ is real and $A > 0$): real λ: (a) $\lambda > 0$ (b) $\lambda < 0$

However when the auxiliary equation has complex roots, sine and cosine terms are introduced into the solution and this means that it will involve oscillations. This is illustrated in the next two examples.

EXAMPLE

Find the particular solution of the differential equation

$$\frac{d^2y}{dx^2} + 16y = 0$$

that satisfies the initial conditions $y = 0$ and $\frac{dy}{dx} = 2$ when $x = 0$.

Solution

The auxiliary equation is

$$\lambda^2 + 16 = 0.$$

and this has roots $\lambda_1 = 4j$ and $\lambda_2 = -4j$.

The general solution of the differential equation is therefore

$$y = A \sin 4x + B \cos 4x.$$

Notice that we go straight to the sine and cosine form.

When $x = 0$, $y = 0$ $\Rightarrow$ $B = 0$.

The solution becomes $y = A \sin 4x$.

Differentiating this solution with respect to x gives

$$\frac{dy}{dx} = 4A \cos 4x.$$

When $x = 0$, $\frac{dy}{dx} = 2$ $\Rightarrow$ $4A = 2$

$\Rightarrow$ $A = \frac{1}{2}.$

The particular solution is $y = \frac{1}{2}\sin 4x$. The diagram shows a graph of the particular solution. You will notice that the solution curve is oscillatory.

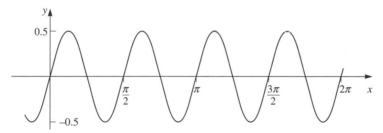

Auxiliary equation with complex roots: the general case

In the previous two examples the auxiliary equations

$$\lambda^2 + 4 = 0 \quad \text{and} \quad \lambda^2 + 16 = 0$$

both had pure imaginary roots, $\pm 2j$ and $\pm 4j$ respectively.

However, many differential equations give rise to auxiliary equations with roots that are part real and part imaginary, as in the next example.

EXAMPLE

Find the particular solution of the differential equation

$$\frac{d^2 y}{dx^2} + 2\frac{dy}{dx} + 5y = 0$$

which satisfies the initial conditions $y = 0$ and $\dfrac{dy}{dx} = 1$ when $x = 0$.

Solution

The auxiliary equation is
$$\lambda^2 + 2\lambda + 5 = 0.$$

Solving for λ, using the quadratic formula (since the equation cannot be factorised):

$$\lambda = \frac{-2 \pm \sqrt{4-20}}{2} = -1 \pm 2j.$$

The general solution is $\quad y = Pe^{(-1+2j)x} + Qe^{(-1-2j)x}$
$$= e^{-x}(Pe^{2jx} + Qe^{-2jx}),$$

and since $Pe^{2jx} + Qe^{-2jx}$ can be written as $A \sin 2x + B \cos 2x$,

$$y = e^{-x}(A \sin 2x + B \cos 2x).$$

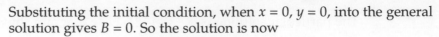

Substituting the initial condition, when $x = 0$, $y = 0$, into the general solution gives $B = 0$. So the solution is now

$$y = Ae^{-x} \sin 2x.$$

Differentiating with respect to x gives

$$\frac{dy}{dx} = -Ae^{-x} \sin 2x + 2Ae^{-x} \cos 2x.$$

Substituting the other initial condition, when $x = 0$, $\frac{dy}{dx} = 1$, into this equation gives

$$2A = 1$$

$$\Rightarrow \qquad A = \frac{1}{2}.$$

The particular solution for these initial conditions is

$$y = \frac{1}{2}e^{-x} \sin 2x$$

The graph of this particular solution is shown below.

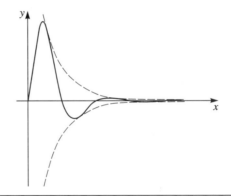

Notice the relationship between the algebraic solution, $y = \frac{1}{2}e^{-x} \sin 2x$, and its graph. The factor e^{-x} tells you that it is subject to exponential decay because of the negative sign in the power; the factor $\sin 2x$ tells you that there is an oscillation. So just by looking at the algebraic expression you can say that the form of the solution is an exponentially decaying oscillation. This form of solution arises frequently when you are modelling real oscillating systems.

For discussion

(i) How would you sketch the curve $y = e^{2x} \cos 3x$ without plotting points or using a graphics calculator?

(ii) What auxiliary equation would give rise to such a solution?

For each of the differential equations in questions 1–5, find the particular solution which corresponds to the given conditions and draw a sketch graph of the solution. Assuming these differential equations are models for real systems, use your graph to describe the motion of each system.

1. $\dfrac{d^2x}{dt^2} + 9x = 0$;

initial conditions $x = 0$ and

$\dfrac{dx}{dt} = 1$ when $t = 0$.

2. $4\dfrac{d^2x}{dt^2} + x = 0$;

initial conditions $x = 4$ and

$\dfrac{dx}{dt} = 0$ when $t = 0$.

3. $\dfrac{d^2x}{dt^2} + 4x = 0$;

boundary conditions $x = 1$ when

$t = \dfrac{\pi}{4}$ and $x = 0$ when $t = \dfrac{\pi}{2}$.

4. $\dfrac{d^2x}{dt^2} + 12x = 0$;

boundary conditions $x = 0$ when

$t = 0$ and $x = 3$ when $t = 1$.

5. $m\dfrac{d^2x}{dt^2} + kx = 0$;

(where m and k are positive constants); initial conditions

$x = 0$ and $\dfrac{dx}{dt} = 1$ when $t = 0$.

6. Find the general solution of each of the following differential equations.

(i) $\dfrac{d^2y}{dx^2} - 4\dfrac{dy}{dx} + 5y = 0$

(ii) $\dfrac{d^2y}{dx^2} - 2\dfrac{dy}{dx} + 5y = 0$

(iii) $\dfrac{d^2x}{dt^2} + 2\dfrac{dx}{dt} + 4x = 0$

(iv) $4\dfrac{d^2x}{dt^2} + 4\dfrac{dx}{dt} + 5x = 0$

7. Find the particular solution of each of the following equations for the conditions given. Use a graphics calculator or computer to graph the solution. Describe the shape of the graph as the independent variable increases.

(i) $\dfrac{d^2y}{dx^2} + 2\dfrac{dy}{dx} + 2y = 0; y = 0$

and $\dfrac{dy}{dx} - 2$ when $x = 0$.

(ii) $4\dfrac{d^2y}{dx^2} - 8\dfrac{dy}{dx} + 5y - 0; y = 2$

and $\dfrac{dy}{dx} = 0$ when $x = 0$.

(iii) $\dfrac{d^2y}{dx^2} + 2\dfrac{dy}{dx} + 5y = 0$;

$y = 0$ when $x = 0$ and

$y = 3$ when $x = \dfrac{\pi}{4}$.

(iv) $\dfrac{d^2x}{dt^2} - 3\dfrac{dx}{dt} + 4x = 0; \dfrac{dx}{dt} = 0$

and $x = 1$ when $t = 0$.

Auxiliary equation with repeated roots

Look at the second order differential equation:

$$\frac{d^2y}{dx^2} - 4\frac{dy}{dx} + 4y = 0.$$

Assuming a solution of the form $y = Ae^{\lambda x}$, you obtain the auxiliary equation

$$\lambda^2 - 4\lambda + 4 = 0$$

$$\Rightarrow \quad (\lambda - 2)^2 = 0.$$

This auxiliary equation is a special case: it has 2 as a repeated root, so it has only one root, $\lambda = 2$. This gives you only one complementary function Ae^{2x}. But since the equation was of second order, you expect the auxiliary equation to have two roots, giving a general solution with two unknown constants.

To find such a solution, try replacing the simple form Ae^{2x} with the more general form $f(x)e^{2x}$, and investigate the form of the function $f(x)$.

Differentiating this solution gives

$$\frac{dy}{dx} = 2e^{2x}f + e^{2x}\frac{df}{dx}.$$

> Using the product rule. Notice that for simplicity $f(x)$ is now written as just f.

Differentiating again gives

$$\frac{d^2y}{dx^2} = 4e^{2x}f + 4e^{2x}\frac{df}{dx} + e^{2x}\frac{d^2f}{dx^2}.$$

Substituting these expressions into the differential equation gives

$$\left(e^{2x}\frac{d^2f}{dx^2} + 4e^{2x}\frac{df}{dx} + 4e^{2x}f\right) - 4\left(e^{2x}\frac{df}{dx} + 2e^{2x}f\right) + 4e^{2x}f = 0.$$

This simplifies to

$$e^{2x}\frac{d^2f}{dx^2} + (4e^{2x} - 4e^{2x})\frac{df}{dx} + (4e^{2x} - 8e^{2x} + 4e^{2x})f = 0.$$

The coefficients of the terms in f and $\frac{df}{dx}$ are zero, so you are left with

$$\frac{d^2f}{dx^2} = 0.$$

Integrating twice gives

$$f = A + Bx$$

where A and B are unknown constants, showing that the general solution has the form

$$y = (A + Bx)e^{2x}.$$

This is an important result. If $\lambda = \alpha$ is a repeated root of the auxiliary equation, you may assume that the general solution of the form

$$y = Ae^{\alpha x} + Bxe^{\alpha x}.$$

EXAMPLE

Find the general solution of the equation

$$\frac{d^2z}{dt^2} - 6\frac{dz}{dt} + 9z = 0.$$

Solution

Assuming a general solution of the form $z = Ae^{\lambda t}$ gives the auxiliary equation

$$\lambda^2 - 6\lambda + 9 = 0$$
$$\Rightarrow \quad (\lambda - 3)^2 = 0$$
$$\Rightarrow \quad \lambda = 3 \text{ (twice)}.$$

The general solution is

$$z = Ae^{3t} + Bte^{3t} = (A + Bt)e^{3t}.$$

Activity

Find the general solution of the differential equation

$$\frac{d^3y}{dx^3} - 3\alpha\frac{d^2y}{dx^2} + 3\alpha^2\frac{dy}{dx} - \alpha^3 y = 0$$

whose auxiliary equation is $(\lambda - \alpha)^3 = 0$.

Exercise 6E

1. Find the general solution of each of the following differential equations.

(i) $\dfrac{d^2y}{dx^2} - 16y = 0$

(ii) $\dfrac{d^2x}{dt^2} + \omega^2 x = 0$

where ω is constant.

(iii) $\dfrac{d^2y}{dx^2} + \dfrac{dy}{dx} - 5y = 0$

(iv) $3\dfrac{d^2x}{dt^2} + \dfrac{dx}{dt} + 2x = 0$

(v) $\dfrac{d^2y}{dx^2} - 8\dfrac{dy}{dx} + 16y = 0$

(vi) $2\dfrac{d^2x}{dt^2} + 4\dfrac{dx}{dt} + x = 0$

(vii) $7\dfrac{d^2y}{dx^2} + 2\dfrac{dy}{dx} = 0$

(viii) $9\dfrac{d^2y}{dx^2} - 12\dfrac{dy}{dx} + 4y - 0$

(ix) $m\dfrac{d^2x}{dt^2} + r\dfrac{dx}{dt} + kx = 0$

(x) $\dfrac{d^4x}{dt^4} - 16x = 0$

2. Show that if a, b and c are positive constants, then both complementary functions of

$$a\frac{d^2x}{dt^2} + b\frac{dx}{dt} + cx = 0$$

approach zero as t tends to infinity.

3. Find the particular solution in each of the following differential equations, then use a graphics calculator or computer to draw a graph of the solution. Assuming that these differential equations are models for real systems, use your graph to describe the motion of each system.

(i) $\dfrac{d^2x}{dt^2} + 8x = 0; \quad x = 1$ and
$\dfrac{dx}{dt} = 1$ when $t = 0.$

(ii) $\dfrac{d^2x}{dt^2} + 5\dfrac{dx}{dt} = 0; \quad x = 0$ and
$\dfrac{dx}{dt} = 4$ when $t = 0.$

(iii) $\dfrac{d^2x}{dt^2} - \dfrac{dx}{dt} + x = 0; \quad x = 1$ and
$\dfrac{dx}{dt} = 0$ when t $= 0.$

4. For each of the following differential equations, find the particular solution that fits the given conditions.

(i) $\dfrac{d^2y}{dx^2} + 2\dfrac{dy}{dx} + y = 0; \quad y = 0$ when
$x = 0$ and $y = 2$ when $x = 1.$

(ii) $\dfrac{d^2x}{dt^2} - 6\dfrac{dx}{dt} + 9x = 0; \quad x = 1$ and
$\dfrac{dx}{dt} = 0$ when $t = 0.$

(iii) $4\dfrac{d^2x}{dt^2} + 4\dfrac{dx}{dt} + x = 0, \quad x = 4$ when
$t = 0$ and $x = 0$ when $t = 2.$

(iv) $\dfrac{d^2y}{dx^2} + 2k\dfrac{dy}{dx} + k^2y = 0; \quad y = 0$ and
$\dfrac{dy}{dx} = 2$ when $x = 0.$

Non-homogeneous equations

You have seen that equations of the form

$$\frac{d^2y}{dx^2} + a\frac{dy}{dx} + by = 0,$$

have general solution

$$y = Au(x) + Bv(x)$$

where u(x) and v(x) may involve exponential and/or trigonometric functions (and possibly a factor x). In modelling real situations, equations like this often arise in which the right hand side is non-zero. You then have a *non-homogeneous* linear equation,

$$\frac{d^2y}{dx^2} + a\frac{dy}{dx} + by = f(x).$$

There are many situations in which these equations arise, such as the parachutist example on p60. One important application is in modelling forced oscillations, and this is discussed in Chapter 8.

As a first step, look at the first order non-homogeneous linear differential equation

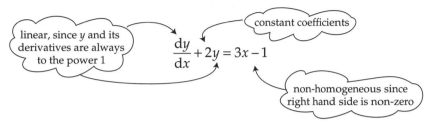

You have seen how to solve this using the integrating factor method. For this equation the integrating factor is e^{2x} so the equation becomes

$$e^{2x}\frac{dy}{dx} + 2e^{2x}y = (3x-1)e^{2x}$$

$$\Rightarrow \quad \frac{d}{dx}(e^{2x}y) = e^{2x}(3x-1)$$

$$\Rightarrow \quad e^{2x}y = \int e^{2x}(3x-1)\,dx$$

Integrating the right hand side by parts gives

$$e^{2x}y = \frac{3}{2}xe^{2x} - \frac{5}{4}e^{2x} + A$$

where A is the constant of integration, so

$$y = Ae^{-2x} + \left(\frac{3}{2}x - \frac{5}{4}\right).$$

Notice that the solution consists of two distinct parts.

- The first part, Ae^{-2x}, contains an arbitrary constant, A, and satisfies the homogeneous equation, $\frac{dy}{dx} + 2y = 0$. It is the complementary function of the equation formed by replacing the right hand side with zero.
- The second part $\left(\frac{3}{2}x - \frac{5}{4}\right)$, includes no arbitrary constants.

This suggests that part of the solution depends only on the left hand side of the differential equation, and part on the full equation. To explore this idea further, look carefully at the next example which has the same left hand side as the previous one, but a different right hand side.

6

EXAMPLE Use the integrating factor method to find the general solution of the first order linear differential equation

$$\frac{dy}{dx} + 2y = e^{3x}.$$

Solution

The integrating factor is again e^{2x} so the general solution is given by

$$e^{2x}\frac{dy}{dx} + 2e^{2x}y = e^{2x}e^{3x}$$

$$\Rightarrow \frac{d}{dx}(ye^{2x}) = e^{5x}$$

$$\Rightarrow ye^{2x} = \int e^{5x}dx$$

$$= \frac{1}{5}e^{5x} + A$$

where A is the constant of integration.

Dividing through by e^{2x}, the general solution is

$$y = \frac{1}{5}e^{3x} + Ae^{-2x}$$

This has the same form, a complementary function Ae^{-2x} and another part, in this case $\frac{1}{5}e^{3x}$, which contains no arbitrary constants.

Now look at the two equations and their solutions.

$$\frac{dy}{dx} + 2y = 3x - 1 \qquad \Rightarrow \qquad y = Ae^{-2x} + \frac{3}{2}x - \frac{5}{4}$$

$$\frac{dy}{dx} + 2y = e^{3x} \qquad \Rightarrow \qquad y = Ae^{-2x} + \frac{1}{5}e^{3x}$$

The complementary function is the same in each case, and could have been found by solving the corresponding homogeneous equation. The other part is the same sort of function as the right hand side of the differential equation. This function is called the *particular integral*.

Activity

Verify that in each case the particular integral is a solution of the full differential equation.

In both cases we solved the differential equations using the integrating factor method, but the form of the solutions suggested that another method might be available, one that involves no integration. The outstanding question is how to find a particular integral.

The first step is to write down a suitable trial function. The form of this trial function will be determined by the right hand side of the differential equation. As an example, take the last differential equation and use ce^{3x} as a trial function. Since you know the solution of this differential equation, you expect c to turn out to be $\frac{1}{5}$.

Substituting $y = ce^{3x}$ and $\dfrac{dy}{dx} = 3ce^{3x}$ into the full differential equation

yields

$$3ce^{3x} + 2ce^{3x} = e^{3x}$$
$$\Rightarrow 5ce^{3x} = e^{3x}$$
$$\Rightarrow \quad c = \frac{1}{5}.$$

The particular integral is $\frac{1}{5}e^{3x}$, as before.

The method that we have developed may now be summarised as follows.

To solve an equation of the form

$$\frac{dy}{dx} + ay = f(x),$$

1. find the complementary function for the associated homogeneous equation

    $$\frac{dy}{dx} + ay = 0;$$

2. find, by using a suitable trial function, a particular integral (that is *any* solution of the full differential equation);
3. the general solution of the equation is now found by adding the complementary function and the particular integral together:

    general solution = complementary function + particular integral.

When applied to first order equations this method may be thought of as being an alternative to the integrating factor method. However it is much more powerful than that because, unlike the integrating factor method, it can be applied to second and higher order equations as well. This is shown in the next example.

EXAMPLE

Given the differential equation

$$\frac{d^2y}{dx^2} - 2\frac{dy}{dx} - 3y = 6e^{2x},$$

(i)   find the complementary function;
(ii) use the function ce^{2x} as a trial function to find a particular integral;
(iii) write down the general solution of the differential equation.

Solution

(i)   The corresponding homogeneous differential equation is

$$\frac{d^2y}{dx^2} - 2\frac{dy}{dx} - 3y = 0$$

The auxiliary equation for this differential equation is

$$\lambda^2 - 2\lambda - 3 = (\lambda - 3)(\lambda + 1) = 0.$$

The complementary function is therefore $Ae^{3x} + Be^{-x}$.

(ii)   If $y = ce^{2x}$, then $\dfrac{dy}{dx} = 2ce^{2x}$ and $\dfrac{d^2y}{dx^2} = 4ce^{2x}$.

Substituting these into the full differential equation:

$$4ce^{2x} - 2(2ce^{2x}) - 3(ce^{2x}) = 6e^{2x}$$
$$\Rightarrow \qquad\qquad\qquad -3ce^{2x} = 6e^{2x}$$
$$\Rightarrow \qquad\qquad\qquad\qquad c = -2.$$

The particular integral is $-2e^{2x}$.

(iii)   The general solution is the complementary function plus the particular integral:

$$y = Ae^{3x} + Be^{-x} - 2e^{2x}.$$

Notice that the general solution given in part (iii) of the last example has the properties you would expect:

- it satisfies the differential equation;
- the number of arbitrary constants is the same as the order of the equation, in this case two.

Finding particular integrals

As you have seen in the examples, to find the complementary function for a non-homogenous equation you use the same technique as for the corresponding homogeneous equation. To find a particular integral you need to establish a trial function whose form depends on the form of f(x).

For discussion

Look at the following three non-homogeneous differential equations.

(i) $\dfrac{d^2y}{dx^2} + 3\dfrac{dy}{dx} + 2y = x^2 + x - 1$

(ii) $\dfrac{d^2y}{dx^2} + 3\dfrac{dy}{dx} + 2y = \sin 4x - \cos 4x$

(iii) $$\frac{d^2y}{dx^2} + 3\frac{dy}{dx} + 2y = 4e^{-0.5x}$$

In each case the complementary function is $Ae^{-x} + Be^{-2x}$.

Suggest a trial function for the particular integral in each case.

You have seen that for a differential equation in which the right hand side is f(x),

- when f(x) is a linear function, the particular integral is also a linear function;
- when f(x) is a trigonometric function, the particular integral is also a trigonometric function;
- when f(x) is an exponential function, the particular integral is also an exponential function.

These observations lead us to use trial functions as given in the following table.

function (x)	trial function
linear function	$lx + m$
polynomial of order n	$a_nx^n + a_{n-1}x^{n-1} + \dots + a_1x + a_0$
trigonometric function involving $\sin px$ and $\cos px$	$l \sin px + m \cos px$
exponential function involving e^{px}	ce^{px}

NOTE

This method finds the simplest form of a particular integral and this is sometimes called the particular integral. Other expressions for a particular integral may be found by adding on multiples of the terms in the complementary function.

The following examples demonstrate how the approach works for a variety of second order equations.

EXAMPLE

Find a particular integral of the differential equation

$$\frac{d^2y}{dx^2} + 3\frac{dy}{dx} + 2y = 4x - 1$$

Solution

Since the function on the right hand side is linear, we use the trial function $y = lx + m$, which gives $\dfrac{dy}{dx} = l$ and $\dfrac{d^2y}{dx^2} = 0$.

Substituting these into the differential equation gives

$$0 + 3l + 2(lx + m) = 4x - 1$$
$$\Rightarrow \quad 2lx + (3l + 2m) = 4x - 1$$

Equating coefficients of x, $2l = 4 \Rightarrow l = 2$

Equating the constant terms, $3l + 2m = -1 \Rightarrow m = -\dfrac{7}{2}$

So a particular integral is $2x - \dfrac{7}{2}$.

EXAMPLE

Find a particular integral of the differential equation

$$\frac{d^2x}{dt^2} - 2\frac{dx}{dt} = 3\cos 2t.$$

Solution

Since the function on the right hand side is trigonometric, we use the trial function

> The angle used in the trial function is the same angle ($2t$) as appears in the function on the right hand side in the differential equation.

$$x = l\sin 2t + m\cos 2t.$$

> Although $\sin 2t$ does not appear on the right hand side of the differential equation, you still need to include it in the trial function.

Differentiating this gives
$$\frac{dx}{dt} = 2l\cos 2t - 2m\sin 2t$$

and
$$\frac{d^2x}{dt^2} = -4l\sin 2t - 4m\cos 2t.$$

Substituting these into the differential equation gives

$$(-4l\sin 2t - 4m\cos 2t) - 2(2l\cos 2t - 2m\sin 2t) \equiv 3\cos 2t$$

$$\Rightarrow \quad (-4l + 4m)\sin 2t + (-4l - 4m)\cos 2t \equiv 3\cos 2t$$

Equating coefficients of $\sin 2t$, $\quad -4l + 4m = 0$

Equating coefficients of $\cos 2t$, $\quad -4l - 4m = 3$

Solving for l and m gives $\quad l = -\frac{3}{8}$ and $m = -\frac{3}{8}$.

The particular integral is therefore
$$-\frac{3}{8}\sin 2t - \frac{3}{8}\cos 2t.$$

> Notice that $\sin 2t$ appears in the particular integral even though it was not present on the right hand side of the differential equation.

Sometimes the right hand side of the differential equation is the sum of different types of function. In such cases your trial function must also be a similar sum, as shown in the next example.

EXAMPLE

Find a particular integral for the differential equation

$$\frac{d^2h}{dt^2} - \frac{dh}{dt} - 6h = 3t + e^{-t}$$

Solution

The trial function needs to have two elements:

to match the $3t$ on the right hand side it needs to contain $lt + m$;

to match the e^{-t} on the right hand side it needs to contain ce^{-t}.

The trial function is therefore $h = lt + m + ce^{-t}$.

Differentiating the trial function, we obtain

$$\frac{dh}{dt} = l - ce^{-t}$$

and $$\frac{d^2h}{dt^2} = ce^{-t}.$$

Substituting these into the differential equation gives

$$ce^{-t} - (l - ce^{-t}) - 6(lt + m + ce^{-t}) \equiv 3t + e^{-t}$$

$$\Rightarrow \quad -6lt - l - 6m - 4ce^{-t} \equiv 3t + e^{-t}$$

Equating coefficients of t: $\quad -6l = 3 \quad \Rightarrow \quad l = -\frac{1}{2}$

Equating constants: $\quad -l - 6m = 0 \Rightarrow m = \frac{1}{12}$

Equating coefficients of e^{-t}: $-4c = 1 \Rightarrow c = -\frac{1}{4}$.

A particular integral is therefore $\quad -\frac{1}{2}t + \frac{1}{12} - \frac{1}{4}e^{-t}$

Special cases

In some differential equations the function on the right hand has the same form as one of the complementary functions. For example, the complementary function of the differential equation

$$\frac{d^2y}{dx^2} - 5\frac{dy}{dx} + 6y = 4e^{3x}$$

is $Ae^{2x} + Be^{3x}$, and e^{3x} occurs on the right hand side. In this situation it is no good using the trial function ce^{3x}, since upon substituting $y = ce^{3x}$,

$\frac{dy}{dx} = 3ce^{3x}$ and $\frac{d^2y}{dx^2} = 9ce^{3x}$ into the differential equation, you obtain

$$9ce^{3x} - 5(3ce^{3x}) + 6(ce^{3x}) = 4e^{3x}$$

$$\Rightarrow \quad 0 = 4e^{3x}$$

and so clearly this trial function does not work.

Instead, $y = cxe^{3x}$ is used as a trial function.

This gives $\dfrac{dy}{dx} = ce^{3x} + 3cxe^{3x}$

and $\dfrac{d^2y}{dx^2} = 6ce^{3x} + 9cxe^{3x}.$

Substituting these in the differential equation gives

$$(6ce^{3x} + 9cxe^{3x}) - 5(ce^{3x} + 3cxe^{3x}) + 6cxe^{3x} \equiv 4e^{3x}$$

$$\Rightarrow \quad ce^{3x} = 4e^{3x}$$

$$\Rightarrow \quad c = 4.$$

A particular integral is $4xe^{3x}$.

This illustrates a general rule. If the function on the right hand side of the differential equation has exactly the same form as one of the complementary functions, you multiply the usual trial function by the independent variable to give a new trial function. In order to recognise these special cases when they arise, it is worth getting into the habit of finding the complementary function before the particular integral.

Activity

Find the values of c and d that would be obtained if the trial function $ce^x + dxe^x$ were to be used to find a particular integral for the equation

$$\frac{d^2y}{dx^2} - 5\frac{dy}{dx} + 6y = e^x.$$

Exercise 6F

Find a particular integral for each of the following differential equations.

1. $\dfrac{d^2y}{dx^2} - 4\dfrac{dy}{dx} + y = -2x + 3$

2. $\dfrac{d^2x}{dt^2} + 4x = t + 2$

3. $4\dfrac{d^2y}{dx^2} - \dfrac{dy}{dx} + 3y = 5x - 1$

4. $\dfrac{d^2x}{dt^2} + \omega^2 x = l$

where ω and l are constants.

5. $\dfrac{d^2v}{dt^2} - 2v = t + 3$

6. $3\dfrac{d^2y}{dx^2} + \dfrac{dy}{dx} = x - 4$

7. $\dfrac{d^2y}{dx^2} + 2\dfrac{dy}{dx} + y = \cos 3x$

8. $\dfrac{d^2y}{dx^2} + 2\dfrac{dy}{dx} + y = \sin 4x$

9. $\dfrac{dv}{dt} - 3v = 2\sin t$

10. $\dfrac{d^2x}{dt^2} + 9x = 3\cos 2t$

11. $\dfrac{d^2x}{dt^2} - 4x = \sin\omega t$

14. $\dfrac{d^2x}{dt^2} + \dfrac{dx}{dt} + 2x = 3e^{-2t}$

12. $\dfrac{d^2x}{dt^2} + 4\dfrac{dx}{dt} + x = \cos t - 2\sin t$

15. $\dfrac{dy}{dx} - 3y = 5e^{2x}$

13. $\dfrac{d^2y}{dx^2} + 4y = e^{-x}$

16. $\dfrac{d^2x}{dt^2} - 5\dfrac{dx}{dt} = 2e^{-3t}$

Finding general and particular solutions

Now that you have seen how to find particular integrals in a variety of situations, you are in a position to find the general solutions of many differential equations. It is then a simple matter to substitute any given conditions into the general solution to find a particular solution. The next example shows how this is done.

EXAMPLE

Find the particular solution of the differential equation

$$\frac{d^2x}{dt^2} + 9x - 13e^{2t} - 18$$

subject to the conditions $x = 1$ and $\dfrac{dx}{dt} = 11$ when $t = 0$.

Solution

The auxiliary equation is $\lambda^2 + 9 = 0$
which has roots $\lambda = -3j$ and $\lambda = 3j$.

The complementary function is therefore

$$x = A \sin 3t + B \cos 3t.$$

To find a particular integral, we use the trial function $x = ce^{2t} + k$. Differentiating this gives

$$\frac{dx}{dt} = 2ce^{2t}$$

and

$$\frac{d^2x}{dt^2} = 4ce^{2t}.$$

Substituting into the differential equation and simplifying:

$$4ce^{2t} + 9(ce^{2t} + k) \equiv 13e^{2t} - 18$$

Equating coefficients of e^{2t}: $\quad 13c = 13 \quad \Rightarrow \quad c = 1$

Equating constants: $\quad 9k = -18 \;\Rightarrow\; k = -2$

A particular integral is $e^{2t} - 2$.

So the general solution of the differential equation is

$$x = A \sin 3t + B \cos 3t + e^{2t} - 2.$$

The initial conditions give the values for the unknown constants A and B:

$x = 1$ when $t = 0 \;\Rightarrow\; B + 1 - 2 = 1 \;\Rightarrow\; B = 2;$

$\dfrac{dx}{dt} = 11$ when $t = 0 \;\Rightarrow\; 3A \cos 3t - 3B \sin 3t + 2e^{2t} = 0$ when $t = 0,$

i.e. $3A + 2 = 11 \;\Rightarrow\; A = 3.$

The particular solution is

$$x = 3 \sin 3t + 2 \cos 3t + e^{2t} - 2.$$

HISTORICAL NOTE

The solution of second and higher order differential equations with constant coefficients was first undertaken by Euler. The treatment of the special case when the auxiliary equation has equal roots is due to d'Alembert.

Jean-le-Rond d'Alembert was born in Paris in 1717. He was abandoned at birth and found by a policeman near the church of Saint Jean-le-Rond, whence his name. As an adult, working at the French Academy he pioneered the development of partial differential equations.

Exercise 6G

1. Find the general solutions of the following differential equations.

(i) $\quad \dfrac{d^2y}{dx^2} + 4y = e^x + e^{2x}$

(ii) $\quad \dfrac{d^2x}{dt^2} - 2\dfrac{dx}{dt} - 3x = 5e^{-2t} + 10e^{4t}$

(iii) $\quad \dfrac{d^2y}{dx^2} + 2y = 3e^x + \sin x$

(iv) $\quad \dfrac{d^2y}{dx^2} + \dfrac{dy}{dx} + y = \cos x + 4$

(v) $\quad \dfrac{d^2y}{dx^2} + \dfrac{dy}{dx} + y = 3x + 2 + e^x$

(vi) $\quad \dfrac{d^2y}{dx^2} + y = \sin x$

(vii) $\quad \dfrac{d^2y}{dx^2} + 3\dfrac{dy}{dx} - 4y = e^x$

(viii) $\quad \dfrac{d^2y}{dx^2} - 4\dfrac{dy}{dx} + 3y = 2e^x + e^{3x}$

(ix) $\quad \dfrac{d^2y}{dx^2} - 6\dfrac{dy}{dx} + 9y = 4e^{3x}$

(x) $\quad \dfrac{d^2y}{dx^2} - 2\dfrac{dy}{dx} = x$

2. Find the particular solution of each of the following differential equations with the given initial conditions.

    (i) $\dfrac{d^2y}{dx^2} - 5\dfrac{dy}{dx} + 6y = 36x;$

          $y = 0$ and $\dfrac{dy}{dx} = -10$ when $x = 0.$

    (ii) $\dfrac{d^2x}{dt^2} + 9x = 20e^{-t};$

          $x = 0$ and $\dfrac{dx}{dt} = 1$ when $t = 0.$

    (iii) $\dfrac{d^2y}{dx^2} + 2\dfrac{dy}{dx} + 5y = 4e^{-x};$

          $y = 0$ and $\dfrac{dy}{dx} = 0$ when $x = 0.$

    (iv) $\dfrac{d^2x}{dt^2} + \dfrac{dx}{dt} - 2x = 20\sin 2t;$

          $x = 2$ and $\dfrac{dx}{dt} = 0$ when $t = 0.$

    (v) $\dfrac{d^2y}{dx^2} + \dfrac{dy}{dx} = x;$

          $y = 1$ and $\dfrac{dy}{dx} = 0$ when $x = 0.$

    (vi) $\dfrac{d^2x}{dt^2} - 3\dfrac{dx}{dt} + 2x = 1 - e^t;$

          $x = 0$ and $\dfrac{dx}{dt} = 1$ when $t = 0.$

    (vii) $\dfrac{d^2y}{dx^2} + 4y = 12\sin 2x;$

          $y = 0$ and $\dfrac{dy}{dx} = 1$ when $x = 0.$

    (viii) $\dfrac{d^2y}{dx^2} + 4\dfrac{dy}{dx} + 5y = \sin 3x;$

          $y = 1$ and $\dfrac{dy}{dx} = 0$ when $x = 0.$

    (ix) $\dfrac{d^2y}{dx^2} + 4\dfrac{dy}{dx} + 4y = 8e^{2x} + 4x;$

          $y = 0$ and $\dfrac{dy}{dx} = 1$ when $x = 0.$

    (x) $\dfrac{d^2y}{dx^2} + 4\dfrac{dy}{dx} + 4y = e^{-2x};$

          $y = 0.5$ and $\dfrac{dy}{dx} = 0$ when $x = 0.$

3. A biological population of size P at time t is growing in an environment which can support a maximum population which is subject to seasonal variation. The growth of the population is described by the first order linear differential equation

    $$\frac{dP}{dt} + P = 100 + 50\sin t.$$

    Find
    (i)   the complementary function of this differential equation,
    (ii) the particular integral,
    (iii) the complete solution given that initially $P = 20,$
    (iv) the mean size of the population after a long time has elapsed,
    (v)   the amplitude of the oscillations of the population.

    [MEI]

4. The second order linear differential equation

    $$y'' + (\tan x)\, y' = e^{-x}\cos x,$$
    $$y(0) = 1,\ y'(0) = \tfrac{1}{2},$$

    is to be solved. It may be reduced to a linear first order equation by the substitution $y' = v.$
    (i)   Write down the first order equation for $v.$
    (ii) Solve this equation to find v as a function of $x.$
    (iii) Hence find y as a function of $x.$
    (iv) Determine the value of y when $x = \tfrac{1}{4}\pi.$

    [The result

    $$\int e^{-x}\cos x\, dx = \tfrac{1}{2}(\sin x - \cos x)e^{-x}$$

    may be assumed.]

    [MEI]

Investigation

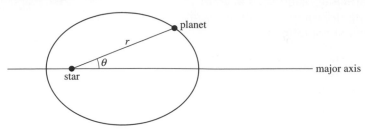

The diagram shows a planet of mass m in an elliptical orbit round a star of mass M. The distance of the planet from the star at time t is r and its displacement vector from the star makes an angle θ with the major axis of the ellipse.

The general orbit under the inverse square law of gravitation is described by the differential equation

$$\frac{\mathrm{d}^2 u}{\mathrm{d}\theta^2} + u = \frac{GM}{h^2}$$

where $u = \dfrac{1}{r}$, G is the universal constant of gravitation, and h is a constant

(h is actually the angular momentum of the planet given by $mr^2 \dfrac{\mathrm{d}\theta}{\mathrm{d}t}$).

Solve this differential equation and investigate the possible solutions. What condition must be satisfied for u to take the value zero? What does this then mean about the nature of the orbit?

KEY POINTS

- A second order linear differential equation with constant coefficients can be written in the form

$$\frac{\mathrm{d}^2 y}{\mathrm{d}x^2} + a\frac{\mathrm{d}y}{\mathrm{d}x} + by = \mathrm{f}(x),$$

where a and b are constants.

- The equation is homogeneous if $\mathrm{f}(x) = 0$. Otherwise it is non-homogenous.

- When you are given a differential equation of the form above, you can immediately write down the auxiliary equation

$$\lambda^2 + a\lambda + b = 0.$$

- Each root of the auxiliary equation determines the form of one of the complementary functions.
- If the auxiliary equation has two real, distinct roots λ_1 and λ_2 then the complementary function of the differential equation is

$$y = Ae^{\lambda_1 x} + Be^{\lambda_2 x}.$$

- If the auxiliary equation has complex roots $\lambda = \alpha + \beta j$ and $\lambda = \alpha - \beta j$, then the complementary function of the differential equation is

$$y = e^{\alpha x}(A \sin \beta x + B \cos \beta x).$$

- If the auxiliary equation has a repeated root $\lambda = \alpha$ then the complementary function of the differential equation is

$$y = e^{\alpha x}(A + Bx) = Ae^{\alpha x} + Bxe^{\alpha x}.$$

- The general solution of the non-homogeneous linear differential equation with constant coefficients

$$\frac{d^2y}{dx^2} + a\frac{dy}{dx} + by = f(x)$$

is the sum of the complementary function and a particular integral.
- The number of unknown constants is the same as the order of the equation.
- A particular integral is any function that satisfies the full equation; it is free of any arbitrary constants.
- To find the particular integral, use the trial function shown in the following table.

function	trial function
linear function	$lx + m$
polynomial of order n	$a_n x^n + a_{n-1}x^{n-1} + \ldots + a_1 x + a_0$
trigonometric function involving $\sin px$ and $\cos px$	$l \sin px + m \cos px$
exponential function involving e^{px}	ce^{px}

- If the trial function for a particular integral is the same as one of the complementary functions, you multiply the trial function by x.

7 **Oscillations**

Human nature oscillates between good and evil

Jowett

The photographs show three situations in which bodies or particles oscillate. List the similarities between the motion of the bodies (or particles) in these situations.

In the next two chapters you will see how the use of differential equations allows you to develop mathematical models which give a very good description of such real situations.

When listing the similarities between different oscillating motions you may have come up with the following.

* The motion follows a regular cyclic (repeating) pattern.
* The body (or particles) move backwards and forwards through a central position.
* The two extreme points of the motion are at equal distances from the central point.
* The velocity of the body (or particles) is greatest at the central position.
* The velocity is instantaneously zero at each extreme of the motion.

These are all true. In fact, remarkably, if you plot the displacement of almost any oscillating body or particle against time, you will find it to be a sine curve.

Notation

Before we can write down any equations we need suitable notation and terminology.

- The central position is usually called O.
- The two extreme points of motion have displacements $\pm a$ from O. The distance a is called the *amplitude* of the motion.
- The number of cycles per second is called the *frequency* of the oscillation, and is usually denoted by v.
- The motion repeats itself after time T. The time interval T is called the *period*: it is the time for one cycle of the motion. Notice that T and v are connected: if the time for one cycle is $\frac{1}{2}$ second, then the frequency is 2 cycles per second. In general

$$T = \frac{1}{v}$$

The SI unit for frequency is the *hertz*. One hertz (1 Hz) is one cycle per second.

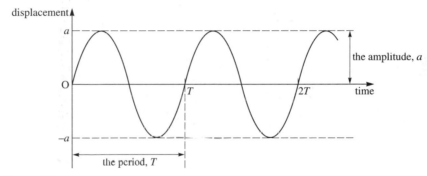

Figure 7.1

Modelling oscillating systems

The first step is to set up mathematical models for the motion of two simple systems, the spring-mass oscillator and the simple pendulum. In each case you will see that applying Newton's Second Law gives rise to a second order differential equation.

The spring-mass oscillator

The spring-mass oscillator consists of an elastic spring attached to a fixed point and at the other end is attached an object. The system is hung vertically as shown in figure 7.2. If the object is displaced vertically from the equilibrium position and then released, the object oscillates along a straight vertical line.

DE

Before you can analyse this motion you
need to make some modelling
assumptions:

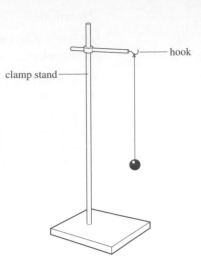

- the effects of friction in the spring and
  air resistance on the object may be
  neglected;
- the spring is light and perfectly elastic:
  this means that the tension in the spring
  is proportional to its extension.

Figure 7.2

The constant of proportionality between the tension and the extension of the
spring is called its *stiffness*, and is denoted by k. The natural length of the
spring is l_0, and the mass of the object is m. You can find the extension, e, of
the spring when the object is in equilibrium by looking at the forces acting on
the object in equilibrium (figure 7.3).

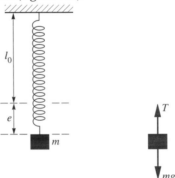

Figure 7.3

There is a downward force mg due to gravity, and a balancing upward
tension in the spring. The tension is given by $T = ke$. Since the system is in
equilibrium

$$ke = mg$$

$$\Rightarrow \qquad e = \frac{mg}{k}.$$

Now look at the system when the spring has a further extension x below the
equilibrium position, so that the total extension is $e + x$ (figure 7.4). The
tension T in the spring is now $T = k(e + x)$.

The object is not in equilibrium, and its acceleration in the downward

direction (the direction of increasing x) is $\dfrac{d^2x}{dt^2}$.

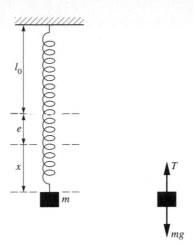

Figure 7.4

Applying Newton's Second Law gives the differential equation

$$m\frac{\mathrm{d}^2x}{\mathrm{d}t^2} = mg - T$$
$$= mg - k(e + x)$$
$$= mg - ke - kx \ .$$

But you have already seen that $mg - ke = 0$.

Substituting this in the differential equation and dividing by m gives the equation of motion

$$\frac{\mathrm{d}^2x}{\mathrm{d}t^2} + \frac{k}{m}x = 0 \qquad \text{or} \qquad \frac{\mathrm{d}^2x}{\mathrm{d}t^2} = -\frac{k}{m}x.$$

This differential equation models the motion of the idealised spring-mass oscillator.

The simple pendulum

A pendulum consists of a small heavy object, the *bob*, suspended from a fixed point by a string. If the bob is displaced by a small angle from its equilibrium position, then it will swing back and forth. Its path will be an arc of a circle.

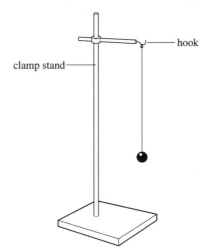

Figure 7.5

As with the spring-mass oscillator, you need to start with some modelling assumptions:

- the bob can be treated as a particle;
- the string is light;
- the string is inelastic (and therefore has constant length);
- the oscillations are small enough for the maximum angle between the string and the vertical to satisfy the approximation $\sin\theta \approx \theta$. (This is accurate to 2 decimal places if the angle is no more than 0.3 radians or 17°.)

This idealised system is called a *simple pendulum*.

Let the mass of the bob be m and the length of the string be l. The bob then swings through a small arc of a circle of radius l. At time t let the angle of the string with the vertical be θ in radians.

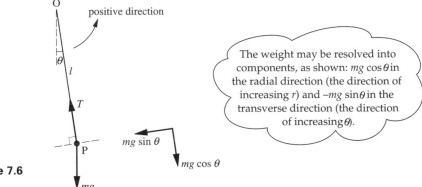

> The weight may be resolved into components, as shown: $mg\cos\theta$ in the radial direction (the direction of increasing r) and $-mg\sin\theta$ in the transverse direction (the direction of increasing θ).

Figure 7.6

Figure 7.6 shows the forces on the object, the weight mg of the bob and the tension T in the string.

The acceleration in the transverse direction is given by $l\dfrac{d^2\theta}{dt^2}$.

Applying Newton's Second Law in the transverse direction gives

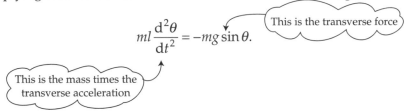

$$ml\frac{d^2\theta}{dt^2} = -mg\sin\theta.$$

> This is the transverse force

> This is the mass times the transverse acceleration

Dividing both sides by m and using the small angle approximation $\sin\theta \approx \theta$, the equation of motion is

$$l\frac{d^2\theta}{dt^2} + g\theta = 0.$$

This can alternatively be written as

$$\frac{d^2\theta}{dt^2} + \frac{g}{l}\theta = 0 \qquad \text{or} \qquad \frac{d^2\theta}{dt^2} = -\frac{g}{l}\theta.$$

This equation models the motion of a simple pendulum.

Simple Harmonic Motion

Look at the differential equations that arose from the two previous examples and you will see something that is both remarkable and exciting. They both have the same form:

$$\frac{d^2x}{dt^2} = -\omega^2 x.$$

In the first case $\omega^2 = \dfrac{k}{m}$, in the second $\omega^2 = \dfrac{g}{l}$. In the second example the variable is called θ rather than x.

The fact that the spring-mass oscillator and the simple pendulum give rise to essentially the same equation of motion shows that the motion is in both cases fundamentally the same. This motion is called *simple harmonic motion* (SHM); it provides a model for many oscillations.

Solving the simple harmonic motion equation

You can use the auxiliary equation method to solve the equation of simple harmonic motion, and this is done below.

The auxiliary equation is

$$\lambda^2 + \omega^2 = 0$$

which has roots $\lambda = \pm j\omega$.

The general solution of the differential equation is then

$$y = A \sin \omega t + B \cos \omega t.$$

The constants of integration, A and B, are unknown at this stage, but if you know suitable initial or boundary conditions you can calculate their values.

Activity

Find the particular solution of the SHM equation given the conditions $x = 0$ when $t = 0$, $x = a$ when $t = \dfrac{\pi}{2\omega}$. Sketch a graph of x against t for this situation.

You should have found that in this case the particular solution was the simple sine expression, $x = a \sin \omega t$, because B turned out to have the value zero. But what happens when the initial conditions are not so convenient?

If the general solution for the motion has the form of a sine curve it should be possible to write $A \sin \omega t + B \cos \omega t$ in the form $a \sin(\omega t + \varepsilon)$, where a is the amplitude of the motion and ε (the Greek letter 'epsilon') is a constant angle. This still contains two unknown constants, but it is more immediately recognisable as a sine function.

DE

To show that this is indeed possible we write

$$A \sin \omega t + B \cos \omega t \equiv a \sin(\omega t + \varepsilon)$$

and expand the right hand side using the compound angle formula. This gives

$$A \sin \omega t + B \cos \omega t \equiv a \sin \omega t \cos \varepsilon + a \cos \omega t \sin \varepsilon.$$

Comparing coefficients of $\sin \omega t$: $A = a \cos \varepsilon$ ①

Comparing coefficients of $\cos \omega t$ $B = a \sin \varepsilon$ ②

Squaring ① and ② and adding them gives

$$A^2 + B^2 = a^2(\cos^2 \varepsilon + \sin^2 \varepsilon) = a^2$$
$$\Rightarrow \quad a = \sqrt{(A^2 + B^2)}$$

Dividing ② by ① gives

$$\tan \varepsilon = \frac{B}{A}.$$

So the general solution of the simple harmonic motion equation can indeed be expressed as a sine function. This tells us a lot about the solution.

- Since the sine function varies between +1 and –1, the solution varies between $+a$ and $-a$.

- Since the sine function is periodic with period 2π, the solution is periodic with period $\dfrac{2\pi}{\omega}$.

- The velocity and acceleration of the motion, found by differentiating the solution with respect to t, are given respectively by $a\omega \cos(\omega t + \varepsilon)$ and $-a\omega^2 \sin(\omega t + \varepsilon)$.

Activity

Given that $x = a \sin(\omega t + \varepsilon)$, sketch the graph of x against t for the following values of ε.

(i) $\varepsilon = 0$ (ii) $\varepsilon = \dfrac{\pi}{4}$ (iii) $\varepsilon = \dfrac{\pi}{2}$

What is the frequency of the oscillations in each case, in terms of ω?

What is the effect of the constant ε?

You should have found in the activity above that in each case the frequency is $\dfrac{\omega}{2\pi}$, and is unaffected by ε. The frequency therefore depends only on ω, which is a constant for a given pendulum, or for a particular spring-mass combination. Notice that ω is not itself the frequency, v; that is given by $v = \dfrac{\omega}{2\pi}$. (The term *angular frequency* is sometimes used to describe ω.)

The effect of ε is simply to shift the sine curve to the left by an amount $\dfrac{\varepsilon}{\omega}$, as shown in figure 7.7.

Changing the magnitude of ε does not affect the frequency of the oscillation or its amplitude. Its effect is just to position the curve correctly relative to the time axis, depending on the point in the cycle at which time is taken to be zero. The quantity ε is called the *phase shift*.

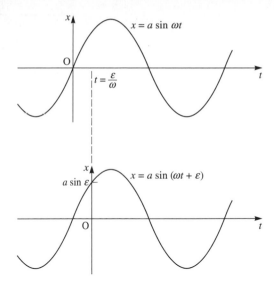

Figure 7.7

Activity

You may sometimes find it more convenient to write the general solution in another equivalent form:

$$y = a \cos(\omega t - \varepsilon_1)$$

Find a and ε_1 in terms of A and B in this case.

EXAMPLE

A spring-mass oscillator consists of a spring of stiffness $20\,\text{Nm}^{-1}$ and natural length $0.5\,\text{m}$ and an object of mass $0.25\,\text{kg}$. The oscillator lies on a smooth horizontal table with one end of the spring attached to a fixed point (as in the diagram). The object is constrained to move along a smooth groove.

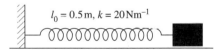

$l_0 = 0.5\,\text{m}, \ k = 20\,\text{Nm}^{-1}$

The object is pulled to the right so that the spring is 0.6 m in length, and the object is released from rest at this point when $t = 0$.

(i)   Form the differential equation for the system and write down the initial conditions.
(ii) Find the particular solution describing the motion.
(iii) Write down the period and amplitude of the motion.
(iv) Sketch a graph of the solution.

Solution

(i)   We assume that the spring is perfectly elastic and has zero mass, and that the effects of friction and air resistance may be neglected. The object is in equilibrium when the spring has its natural length. At time t (in seconds), let the extension of the spring be x metres. (At this point the object is a distance x to the right of its equilibrium position.)

DE

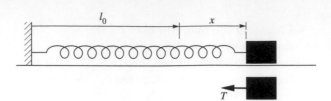

The acceleration of the object at time t is $\dfrac{d^2x}{dt^2}$ in the positive horizontal direction. The only horizontal force is the tension, T newtons, in the spring. The tension is given by

$$T = kx = 20x,$$

Applying Newton's Second Law $F = ma$ gives

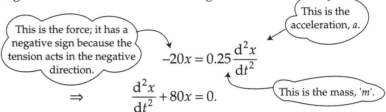

This is the force; it has a negative sign because the tension acts in the negative direction.

This is the acceleration, a.

$$-20x = 0.25\frac{d^2x}{dt^2}$$

This is the mass, 'm'.

$$\Rightarrow \qquad \frac{d^2x}{dt^2} + 80x = 0.$$

This is the differential equation of motion of the system.

Initially the object is at rest, so $\dfrac{dx}{dt} = 0$ when $t = 0$. The extension of the spring is initially 0.1 m, so $x = 0.1$ when $t = 0$. These are the initial conditions.

(ii) The equation of motion is

$$\frac{d^2x}{dt^2} + 80x = 0,$$

which has the form is $\quad x = A\sin\omega t + B\cos\omega t \quad$ where $\omega = \sqrt{80}$.

$$\frac{d^2x}{dt^2} = \omega^2 x.$$

Substituting the initial conditions into the general solution:

When $t = 0$, $x = 0.1$ $\qquad \Rightarrow \qquad 0.1 = B$.

When $t = 0$, $\dfrac{dx}{dt} = 0$ $\qquad \Rightarrow \qquad 0 = A\omega$, and so $A = 0$.

The particular solution is

$$x = 0.1\cos\sqrt{80}t$$

(iii) The amplitude of the motion is 0.1 m and the period is

$$\frac{2\pi}{\omega} = \frac{2\pi}{\sqrt{80}} \approx 0.7 \text{ seconds.}$$

(iv) The graph of displacement x against time t is as shown.

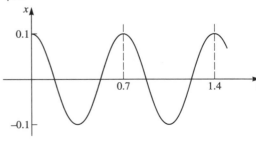

Activity

Starting with the general solution of the SHM equation, $x = a \sin(\omega t + \varepsilon)$, show that when the displacement is x, the velocity and acceleration are given by $v = \pm \omega \sqrt{(a^2 - x^2)}$ and $a = -\omega^2 x$.

Exercise 7A

1. A spring-mass oscillator consists of a spring of stiffness $32\,\text{Nm}^{-1}$, natural length $0.5\,\text{m}$ and an object of mass $0.5\,\text{kg}$. The oscillator lies on a smooth horizontal table with one end of the spring attached to a fixed point. The object is constrained to move in a straight line, and friction can be taken to be negligible.

Initially the object is at rest, and the spring is $0.6\,\text{m}$ in length.
   (i)   Formulate the differential equation for the system and write down the initial conditions in this case.
   (ii) Find the particular solution describing the motion.
   (III) Write down the period and amplitude of the motion.
   (iv) Sketch a graph of the solution.

2. A spring-mass oscillator consists of a spring of natural length $0.2\,\text{m}$ and stiffness $25\,\text{Nm}^{-1}$ and an object of mass $0.4\,\text{kg}$. One end of the spring is attached to a fixed point and the system hangs vertically.
   (i)   Find the length of the spring when the system is in equilibrium.

The object is pulled down $10\,\text{cm}$ from its equilibrium position and released.
   (ii) Formulate the differential equation for the system and write down the initial conditions in this case.
   (iii) Find the particular solution describing the motion.
   (iv) Write down the period and amplitude of the motion.
   (v)   Describe the motion of the object.

3. An object of mass 200 grams is attached to a light spring of natural length $40\,\text{cm}$ and stiffness $50\,\text{Nm}^{-1}$. The object is allowed to hang vertically in equilibrium.
   (i)   Find the extension of the spring in this position.

The spring is now pulled down by a further $3\,\text{cm}$ and released from rest.
   (ii) Find the length of the spring as a function of time.

4. Two springs have the same natural length l but have stiffnesses $75\,\text{Nm}^{-1}$ and $50\,\text{Nm}^{-1}$. An object of mass $0.5\,\text{kg}$ is attached to both springs and their other ends are attached to two points which are a distance $d\,\text{m}$ apart on a smooth horizontal table (where $d > 2l$). Find the period of small oscillations of this system.

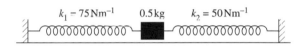

$k_1 = 75\,\text{Nm}^{-1}$ $\qquad$ $0.5\,\text{kg}$ $\qquad$ $k_2 = 50\,\text{Nm}^{-1}$

5. A simple pendulum of length $3\,\text{m}$ has a bob of mass $0.2\,\text{kg}$. It is hanging vertically when it is set in motion by a single, sharp sideways blow to the bob which causes the pendulum to oscillate with an angular amplitude of $5°$.
   (i)   State the amplitude in radians.
   (ii) Calculate the approximate period of the pendulum.
   (iii) The motion of the pendulum can be modelled by the equation
   $$\theta = a \sin(\omega t + \varepsilon).$$
   Write down the values of a, ω and ε.
   (iv) Calculate the maximum speed of the pendulum bob according to this model.

6. A pendulum has a bob of mass $0.25\,\text{kg}$. The period of the pendulum is 2 seconds. The amplitude of its swing is $3°$.
   (i)   Find the length of the pendulum.
   (ii) State the effect (if any) on the period of the pendulum of:

(a) making the mass of the bob 0.75 kg;
(b) doubling the length of the pendulum;
(c) halving the amplitude of its swing;
(d) moving it to the moon where the acceleration due to gravity is $\frac{1}{6}g$.

7. A pendulum consists of a bob of mass 0.2 kg attached to a light inelastic string of length 4 m.

(i) Formulate the equation of motion of the bob, stating any assumptions involved.

(ii) Calculate the period of small oscillations.

8. A loaded test tube of total mass m floats in water and is in equilibrium when a length l is submerged as shown. The upward force exerted at any instant by the water on the tube is F.

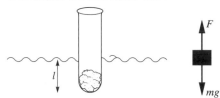

(i) Why is F equal to mg in the equilibrium position? Given that F is directly proportional to the submerged length, find the constant of proportionality in terms of l, m and g.

The test tube is now pushed down a small amount and released.

(ii) Find F when the bottom of the tube is a distance x below its equilibrium position, and use Newton's Second Law to write down the equation of motion.

(iii) Find the period of the motion.

9. A particle of mass 0.02 kg is performing simple harmonic motion with centre O, period 6 s and amplitude 2 m. Find, correct to 2 significant figures,

(i) the maximum speed V of the particle;

(ii) the distance of the particle from O when its speed is $0.9\,V$;

(iii) the magnitude of the force acting on the particle when it is 1 m from O.

Investigation

A *seconds pendulum* is one that swings through its complete arc in 1 second, so that it has a period of 2 seconds.

Set up a seconds pendulum experimentally. What is the length of the pendulum?

Is this what you would have predicted from your knowledge of the SHM model?

Improving the model

Simple harmonic motion has constant amplitude and goes on for ever. For many real oscillating systems, SHM is not a very good model: usually the amplitude of the oscillations gradually decreases, and the motion dies away.

When you find that a model is unsatisfactory, you need to look again at your assumptions. In this case, the assumption that you need to question is that the effects of air resistance and friction can be neglected. Real oscillating systems are almost always *damped*: that is they are affected to some degree by the resistive forces of friction and/or air resistance. They perform *damped oscillation*.

In many systems the damping force is proportional to the speed of the object. This is often represented on a diagram by a device called a linear dashpot, as shown in figure 7.8.

Figure 7.8

A dashpot exerts a force on the system which is proportional to the rate at which it is being extended or compressed, and which acts in the direction opposite to that of the motion. This is illustrated in figure 7.9

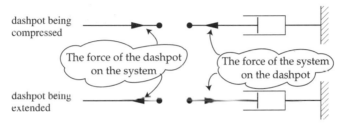

Figure 7.9

The force R that the dashpot exerts on the system at time t is given by

$$R = r\frac{dL}{dt}$$

where the constant of proportionality r is called the *dashpot constant* (or the *damping constant*). The amount of travel still left in the dashpot (see figure 7.10) is denoted by L.

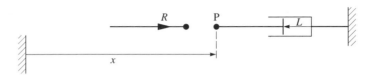

Figure 7.10

It is important to be clear in your mind about the direction of the force R and the signs involved. Look at the point P on the moving part of the dashpot.

- When P is moving from right to left, L is increasing; $\dfrac{dL}{dt}$ is positive and the force is in the same direction as that marked for R in figure 7.10. The dashpot is opposing the right to left motion.
- When P is moving from left to right, L is decreasing, $\dfrac{dL}{dt}$ is negative and the force is in the opposite direction to that marked for R in figure 7.10. The dashpot is now opposing the left to right motion.

Thus the sign of the dashpot force looks after itself as the motion changes.

However you will not usually be interested in the quantity L so much as the distance of the point P from some fixed point of the system. This distance is shown as x in figure 7.10. All the systems that you will meet in this book are set up so that as x increases, L decreases, and vice versa, so

$$\frac{dx}{dt} = -\frac{dL}{dt}$$

Consequently the force that the dashpot exerts on the system is given by

$$R = -r\frac{dx}{dt}$$

in the direction of increasing x.

NOTE

Here we are using a dashpot as a symbol to represent damping effects but a dashpot is actually a real device that is often used in man-made systems. It is used when a certain (predictable) amount of damping, over and above that provided by friction and air resistance, is desired. For example, a car's shock-absorber includes a linear dashpot, as shown in the photograph.

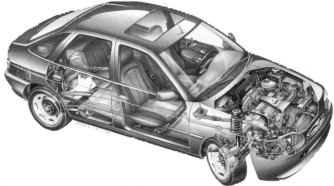

Such a dashpot consists of a cylinder containing a viscous liquid, often oil. When the cylinder is compressed or extended a disc moves along the cylinder, and is opposed by a force which is proportional to the speed of compression or extension. The constant of proportionality depends on the viscosity of the liquid.

EXAMPLE

A simple oscillating system is being modelled as a damped spring-mass oscillator, in which an object of mass $2\,\mathrm{kg}$ is attached to fixed points by a spring of natural length $0.5\,\mathrm{m}$, stiffness $20\,\mathrm{Nm}^{-1}$ and by a dashpot of constant $12\,\mathrm{Nm}^{-1}\mathrm{s}$. The spring–mass–dashpot system lies on a smooth horizontal surface, as shown in the diagram.

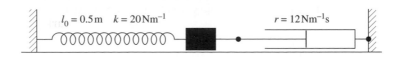

$l_0 = 0.5\,\text{m} \quad k = 20\,\text{Nm}^{-1}$ $\qquad\qquad r = 12\,\text{Nm}^{-1}\text{s}$

(i) Formulate the differential equation of motion for this system.

The system is released from rest when the spring length is $0.6\,\text{m}$.

(ii) Find the particular solution of the differential equation that models this situation.

Solution

(i) The diagram below shows the spring–mass–dashpot system at some general time t (seconds), when the extension of the spring is x. The horizontal forces are the tension in the spring, T, and the damping force R.

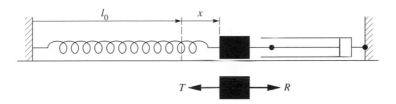

The tension in the spring is $T = kx = 20x$.

The dashpot force is $R = -r\dfrac{dx}{dt} = -12\dfrac{dx}{dt}$.

Applying Newton's Second Law $F = ma$ at any instant gives

$\left(\text{This side is }'ma'\right) \quad 2\dfrac{d^2x}{dt^2} = -12\dfrac{dx}{dt} - 20x \quad \left(\begin{array}{c}\text{This side is }'F' \\ \text{where } F = R - T\end{array}\right)$

Dividing both sides by 2, and rearranging, we obtain the equation of motion of the spring–mass–dashpot system:

$$\dfrac{d^2x}{dt^2} + 6\dfrac{dx}{dt} + 10x = 0. \quad \left(\begin{array}{c}\text{Notice that we again have a linear} \\ \text{homogeneous second order} \\ \text{equation with constant coefficients.}\end{array}\right)$$

(ii) The auxiliary equation for the differential equation is

$$\lambda^2 + 6\lambda + 10 = 0$$

$$\Rightarrow \quad \lambda = \dfrac{-6 \pm \sqrt{(6^2 - 40)}}{2} = -3 \pm j$$

The general solution of the differential equation is $x = ae^{-3t}\sin(t + \varepsilon)$.

At the start of the motion, the length of the spring is $0.6\,\text{m}$ and the object is at rest, so the initial conditions are $x = 0.1$ and $\dfrac{dx}{dt} = 0$ when $t = 0$.

When $t = 0$, $x = 0.1$ $\Rightarrow$ $a \sin \varepsilon = 0.1$. ①

By differentiating the general solution we get

$$\frac{dx}{dt} = -3e^{-3t}a\sin(t+\varepsilon) + ae^{-3t}\cos(t+\varepsilon).$$

When $t = 0$, $\dfrac{dx}{dt} = 0$ $\Rightarrow$ $0 = -3a\sin\varepsilon + a\cos\varepsilon$

$\Rightarrow$ $\tan\varepsilon = \frac{1}{3}$.

From ① we see that $\sin\varepsilon = \dfrac{0.1}{a}$, which is positive, and so ε must be

an angle in the first quadrant.

$\Rightarrow$ $\varepsilon = 0.322$ (radians) and $a = \dfrac{0.1}{\sin 0.322} = 0.316$.

The particular solution in this case is

$$x = 0.316e^{-3t}\sin(t + 0.322).$$

The initial amplitude of the motion is $0.316\,\text{m}$ (to 3 decimal places) and the period in seconds is $\dfrac{2\pi}{1} = 2\pi$.

The amplitude decays exponentially. In this case the oscillation decays very quickly. The diagram below shows the graph of a typical damped oscillation.

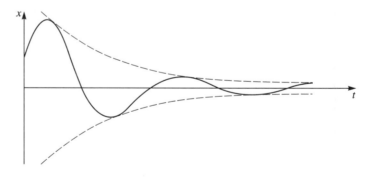

There are many real situations where the oscillations decrease gradually in amplitude like this. Oscillations of this type are often called *lightly damped* or *underdamped*.

The general equation for damped oscillations

The differential equation of motion in the spring–mass–dashpot system above is an example of the general differential equation of a linearly damped system:

$$\frac{d^2y}{dt^2} + \alpha \frac{dy}{dt} + \omega^2 y = 0,$$

where α and ω are positive constants. For a spring–mass–dashpot system, $\alpha = \dfrac{r}{m}$ where r is the dashpot constant, and $\omega = \sqrt{\dfrac{k}{m}}$ where k is the stiffness of the spring. The quantity $\dfrac{\omega}{2\pi}$ is called the *natural frequency* of the system: it is the same whether the system is damped or undamped.

The solution of this differential equation can give several different types of motion, depending on the relative sizes of the parameters α and ω. The auxiliary equation is

$$\lambda^2 + \alpha\lambda + \omega^2 = 0.$$

This has two solutions,

$$\lambda_1 = \frac{-\alpha + \sqrt{(\alpha^2 - 4\omega^2)}}{2} \qquad \text{and} \qquad \lambda_2 = \frac{-\alpha - \sqrt{(\alpha^2 - 4\omega^2)}}{2}.$$

The *discriminant*, $\alpha^2 - 4\omega^2$, determines the nature of the solution. There are three possibilities, as follows.

- *Overdamping*: $\alpha^2 - 4\omega^2$ is positive, and the system does not oscillate (Figure 7.11);

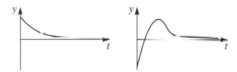

Figure 7.11

- *Underdamping*: $\alpha^2 - 4\omega^2$ is negative, and oscillations occur (Figure 7.12);

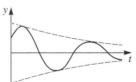

Figure 7.12

- *Critical damping*: $\alpha^2 - 4\omega^2 = 0$ (Figure 7.13).

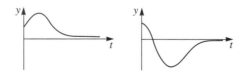

Figure 7.13

Critical damping is the borderline between overdamping and underdamping. It is not obvious in a physical situation when damping is critical, since the pattern of motion for critical damping can be very similar to that in the overdamped case.

For discussion

In many systems damping is desirable or even necessary. For each of the following systems, decide whether damping is desirable and if so whether overdamping, underdamping or critical damping is preferable.

(i)   A clock pendulum   (ii) A car suspension system
(iii) A set of kitchen scales (iv) A robotic welding arm.

Experiment Set up a spring–mass–dashpot system as shown in the diagram.

Displace the object through a small distance then release it. Describe the subsequent motion of the object.

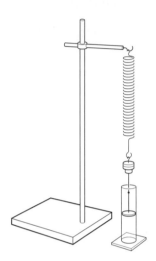

Repeat with objects of different mass and if possible with different liquids in the dashpot. Compare the types of motion that occur.

Now use the apparatus to investigate the following questions.

- Do oscillations always occur?
- If oscillations do occur, is their amplitude constant?
- If oscillations occur, is their period constant?
- If oscillations occur, does their rate of decay depend on the mass?

Design and carry out an experiment which will estimate the damping constant of the dashpot for two liquids such as water and oil.

Exercise 7B

1. A spring–mass oscillator consists of a spring with stiffness $20\,\mathrm{Nm^{-1}}$ and an object of mass $0.25\,\mathrm{kg}$. The object is pulled down $10\,\mathrm{cm}$ from its equilibrium position and released.

(i)   Formulate the differential equation and write down the initial conditions for the system, stating any modelling assumptions you make.

(ii) Find the particular solution describing the motion.

(iii) Sketch a graph of the solution.

A linear damping device is now introduced to the system.

(iv) Calculate the value of the damping constant if the system is to be critically damped.

(v)   If the damping constant does not change, describe the motion of the system if

(a)   the mass of the object is increased to $0.3\,\mathrm{kg}$

(b)   the mass of the object is decreased to $0.2\,\mathrm{kg}$.

2. The pointer on a set of kitchen scales oscillates before settling down at its final reading.

If x is the reading at time t then the oscillation of the pointer is modelled by the differential equation

$$\frac{d^2x}{dt^2} + 3\frac{dx}{dt} + 10x = 0.5$$

(i) Find the general solution of the equation for x.

(ii) Given that $x = 0.1$ and $\frac{dx}{dt} = 0$ when $t = 0$, find the particular solution. At what reading will the pointer settle?

(iii) What length of time will elapse before the amplitude of the oscillations of the pointer is less than 20% of the final value of x?

3. An electrical circuit consists of a 0.2 henry inductor, a 1 ohm resistor and a 0.8 farad capacitor in series. The charge q coulombs on the capacitor is modelled by the differential equation

$$0.2\frac{d^2q}{dt^2} + \frac{dq}{dt} + \frac{1}{0.8}q = 0.$$

Initially q is 2 and $\frac{dq}{dt}$ (the current in ampères) is 4.

(i) Find an equation for the charge as a function of time.

(ii) Sketch the graphs of charge and current against time. Describe how the charge and current change.

(iii) What is the charge on the capacitor and the current in the circuit after a long period of time?

4. The temperature of a chemical undergoing a reaction is modelled by the differential equation

$$2\frac{d^2T}{dt^2} + \frac{dT}{dt} = 0$$

where T is the temperature in °C and t is the time in minutes.

For a particular experiment, the temperature is initially 50 °C, and it is 45 °C one minute later.

(i) Find an expression for the temperature T at any time.

(ii) What will the temperature be after two minutes?

(iii) Sketch a graph of T against t.

(iv) What is the steady state temperature?

5. The angular displacement from its equilibrium position of a swing door fitted with a damping device is modelled by the differential equation

$$\frac{d^2\theta}{dt^2} + 4\frac{d\theta}{dt} + 5\theta = 0.$$

The door starts from rest at an angle of $\frac{\pi}{4}$ from its equilibrium position.

(i) Find the general solution of the differential equation.

(ii) Find the particular solution for the given initial conditions.

(iii) Sketch a graph of the particular solution, and hence describe the motion of the door.

(iv) What does your model predict as t becomes large?

6. A buffer on a railway truck consists of a spring and a damping device in parallel. The truck has mass 10 tonnes and collides with a rigid stop at a speed of $5\,\mathrm{ms^{-1}}$. The spring has stiffness $10^6\,\mathrm{Nm^{-1}}$ and the damping constant is $2 \times 10^5\,\mathrm{Nm^{-1}s}$.

Assume that the buffer on the truck remains in contact with the rigid stop throughout the subsequent motion of the truck.

(i) Formulate and solve a differential equation to model the motion of the railway truck after the collision.

(ii) Sketch a graph of the solution, and from your graph describe the motion of the truck.

7. A simple model of a delicate set of laboratory scales comprises a spring of stiffness $5\,\mathrm{Nm^{-1}}$ and a damping device with constant $7\,\mathrm{Nm^{-1}s}$. An object of mass m is placed carefully on the scales when the spring is unstretched. Take g to be $10\,\mathrm{ms^{-2}}$, and let x m be the displacement from the initial position.

   (i) Formulate and solve a differential equation to model the motion of the object in each of the cases

       (a) $m = 2$ (b) $m = 2.45$

       (c) $m = 4$.

   (ii) Sketch a graph of the solution for objects of each value of m. Describe the motion in each case.

8. A hydrometer is an instrument used to measure the densities of liquids. It works on the principle that a floating body experiences an upthrust equal in magnitude to the weight of liquid displaced by that part of the body which is submerged.

   A particular hydrometer consists of a light uniform cylindrical rod of radius a and length b with a small mass m countersunk into one end. In a liquid the hydrometer floats with its axis vertical and the amount by which it protrudes above the surface gives a measure of the density of the liquid.

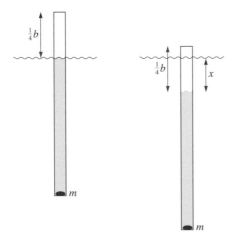

Such a hydrometer floats in equilibrium in a liquid of density ρ with one quarter of its length above the surface.

(i) Find an expression for the density ρ in terms of a, b and m.

The hydrometer is pushed vertically downwards and then released. The resistance to motion when the speed is v is given by $2mkv$ where k is a constant. At time t the displacement of the equilibrium level below the surface is denoted by x. The differential equation governing the motion is

$$m\frac{\mathrm{d}^2x}{\mathrm{d}t^2} = -2mk\frac{\mathrm{d}x}{\mathrm{d}t} - \frac{4mg}{3b}x.$$

(ii) Explain the physical significance of each of the three terms.

(iii) Show that for the hydrometer to perform damped simple harmonic motion

$$\frac{4g}{3b} > k^2.$$

Initially the hydrometer is pushed vertically downwards until just submerged and then released from rest.

(iv) Denoting $\dfrac{4g}{3b} - k^2$ by ω^2, solve the differential equation governing the motion and express the solution as an equation for x in terms of t, b, k and ω.

(v) Describe the motion of the hydrometer by sketching the graph of x against t.

Four complete oscillations after release, one eighth of the hydrometer is above the surface.

(vi) Show that the periodic time T of the oscillations is given by

$$T = \frac{\ln 2}{4k}.$$

[MEI]

Exercise 7B continued

9. Many systems can be modelled by the non-homogeneous differential equation

    $$\frac{d^2x}{dt^2} + \omega^2 x = c \qquad ①$$

    where c is a constant.

    (i) Show that introducing the new variable y, where $y = x - \dfrac{c}{\omega^2}$, gives the homogeneous differential equation

    $$\frac{d^2y}{dt^2} + \omega^2 y = 0.$$

    (ii) Hence write down the general solution of equation ①. Give a physical interpretation of the quantity $\dfrac{c}{\omega^2}$.

    (iii) Use similar ideas to write down the general solution of the equation

    $$\frac{d^2x}{dt^2} + r\frac{dx}{dt} + \omega^2 x = c$$

    where $r > 0$ and c is a constant.

KEY POINTS

-   Motion for which the differential equation is $\dfrac{d^2y}{dt^2} + \omega^2 y = 0$ is called *simple harmonic motion* (SHM).
-   A spring–mass oscillator and a simple pendulum both perform SHM if friction and air resistance are negligible.
-   The period of this SHM is $T = \dfrac{2\pi}{\omega}$.
-   The angular frequency of the SHM is ω and the frequency is $\nu = \dfrac{\omega}{2\pi}$.
-   Linear damping is provided by a linear dashpot.
-   Motion for which the differential equation is

    $$\frac{d^2y}{dt^2} + \alpha\frac{dy}{dt} + \omega^2 y = 0 \text{ (for } \alpha > 0)$$

    is called *damped harmonic motion*. A spring–mass–dashpot system performs damped harmonic motion.

KEY POINTS *continued*

- The following table shows the features of damped harmonic motion.

$\alpha^2 - 4\omega^2$	Solution	Type of damping
$\alpha = 0$	$y = A \sin \omega t + B \cos \omega t$ or $y = a \sin (\omega t + \varepsilon)$	no damping
positive $(\alpha > 2\omega)$	$y = Ae^{\lambda_1 t} + Be^{\lambda_2 t}$	overdamping
zero $(\alpha = 2\omega)$	$y = (A + Bt)e^{-\frac{\alpha t}{2}}$	critical damping
negative $(\alpha < 2\omega)$	$y = ae^{-\frac{\alpha t}{2}} \sin(\rho t + \varepsilon)$, where $\rho = \frac{1}{2}\sqrt{(4\omega^2 - \alpha^2)}$	underdamping

8 Forced oscillations

One cannot escape the feeling that these mathematical formulae have an independent existence and an intelligence of their own, that they are wiser than we are, wiser even than their discoverers, that we get more out of them than was originally put into them.

Heinrich Hertz

Cleo Laine can break a wine glass by singing at it.

(i)   **How is it possible?**
(ii) **Can you do it?**

All structures have natural frequencies of vibration. If an external agent causes them to vibrate at or near one of these frequencies, large oscillations are seen to build up. This phenomenon is called *resonance*. Over the years there have been several well-known instances of resonance occurring with dramatic and destructive effect.

- In 1963 a cooling tower at the Ferrybridge power station near Leeds collapsed. It had been caused to vibrate violently by vortices formed when the wind blew past a row of towers in front of it.
- In 1940 a wind of moderate speed caused large oscillations in the Takoma Narrows bridge in Washington, USA. The bridge eventually collapsed into the river below.
- In 1831 a column of soldiers marching over the Broughton suspension bridge, near Manchester, set up a forced vibration which had a frequency close to one of the natural frequencies of the bridge. The bridge collapsed. Since this disaster, it has become normal for troops to break step when crossing a bridge.
- In biblical times, after camping outside the walled town of Jericho for seven days, Joshua gave orders which brought the siege to an end in a curious way:

So they blew trumpets, and when the army heard the trumpet sound, they raised a great shout, and down came the walls.

Could this, too, have been the effect of forced vibrations?

These instances show how important it is for engineers to be able to predict the natural frequencies of any structures they are designing, and the frequencies of any possible forcing agents. Nor is this a matter for civil engineers alone: engines which run at particular frequencies cause ships and aircraft to shake themselves to pieces.

Modelling forced vibrations – the undamped case

In order to understand the mathematics of forced oscillations, including resonance, we look not at one of the complicated structures in the examples above, but at the simplest suitable case, that of an object hanging on a light, perfectly elastic spring, without damping. (The case in which both forcing and damping occur is considered in the second half of this chapter, on page 122.)

The top end A of the spring is forced to vibrate so that its displacement at time t is $y = a \sin \Omega t$. (This can be achieved experimentally, to a reasonable approximation, by attaching the supporting string over a pulley to a rotating cam, as shown on page 120.)

If the natural length of the spring is l_0, the stiffness of the spring is k and the object has mass m, then in equilibrium (figure 8.1), point A coincides with O, the object is at rest and the extension of the spring is $e = \dfrac{mg}{k}$. At a general time

t during the forced motion, the extension of the spring below the equilibrium position is denoted by x (figure 8.2).

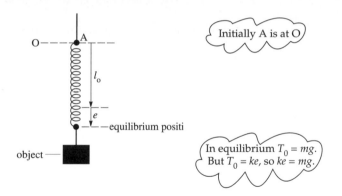

Figure 8.1

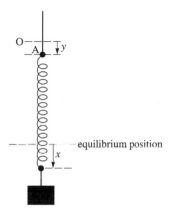

Figure 8.2

There are two forces acting on the object, the force of gravity mg and the tension T. The acceleration of the object is $\dfrac{d^2x}{dt^2}$. Applying Newton's Second Law gives

$$m\frac{d^2x}{dt^2} = mg - T.$$

The extension of the spring is $(e + x - y)$, so the tension in the spring is

$$T = k(e + x - y).$$

The equation of motion is therefore

$$m\frac{d^2x}{dt^2} = mg - k(e + x - y).$$

Expanding the right hand side and recalling that $mg - ke = 0$ (figure 8.1), this becomes

$$m\frac{d^2x}{dt^2} + kx = ky.$$

Dividing both sides by m and putting $\omega^2 = \dfrac{k}{m}$,

$$\frac{d^2x}{dt^2} + \omega^2 x = \omega^2 y \quad \text{or} \quad \frac{d^2x}{dt^2} + \omega^2 x = w^2 f(t) \qquad \text{①}$$

In the system we have described, $\quad y = f(t) = a \sin \Omega t$. This is *forced harmonic motion*, and $\dfrac{\Omega}{2\pi}$ is called the forcing frequency. The differential equation of motion may be written

$$\frac{d^2x}{dt^2} + \omega^2 x = a\omega^2 \sin \Omega t. \qquad \text{②}$$

The complementary function is given by

$$A \sin \omega t + B \cos \omega t.$$

For the particular integral, try $x = p \sin \Omega t + q \cos \Omega t$. This gives

$$\frac{dx}{dt} = p\Omega \cos \Omega t - q\Omega \sin \Omega t$$

and $\quad\dfrac{d^2x}{dt^2} = -p\Omega^2 \sin \Omega t - q\Omega^2 \cos \Omega t.$

> Normally we would use l and m in the trial function, but in this example p and q are used instead because l and m are representing length and mass

Substituting these in the differential equation of motion gives

$$-p\Omega^2 \sin \Omega t - q\Omega^2 \cos \Omega t + p\omega^2 \sin \Omega t + q\omega^2 \cos \Omega t \equiv a\omega^2 \sin \Omega t.$$

Equating coefficients:

$\sin \Omega t$: $\qquad\qquad -p\Omega^2 + p\omega^2 = a\omega^2.$

$\cos \Omega t$: $\qquad\qquad -q\Omega^2 + q\omega^2 = 0.$

Assuming $\Omega \neq \omega$, this gives $\quad p = \dfrac{a\omega^2}{\omega^2 - \Omega^2}\quad$ and $q = 0$.

The particular integral is therefore $\quad\dfrac{a\omega^2}{\omega^2 - \Omega^2} \sin \Omega t.$

The general solution of the differential equation ② is therefore

$$x = A \sin \omega t + B \cos \omega t + \frac{a\omega^2}{\omega^2 - \Omega^2} \sin \Omega t \qquad (\Omega \neq \omega). \qquad \text{③}$$

For discussion

What is the physical significance of the terms

(i) $A \sin \omega t + B \cos \omega t$ $\qquad$ (ii) $\qquad \dfrac{a\omega^2}{\omega^2 - \Omega^2} \sin \Omega t$

in the general solution above?

What would happen to the solution if Ω were increased steadily from zero towards ω?

Describing forced oscillations

From your discussions you should have made the following deductions.

- The terms $A \sin \omega t + B \cos \omega t$ represent the natural or free oscillations of the system, as they would occur if the cam were not rotating. The natural frequency of the system is $\dfrac{\omega}{2\pi}$. You can see that these

  oscillations are as described for the spring–mass oscillator in Chapter 7.

- The term $\dfrac{a\omega^2}{\omega^2 - \Omega^2} \sin \Omega t$ represents the oscillations caused by the

  rotating cam.

- As the value of Ω approaches that of ω, the quantity $\omega^2 - \Omega^2$ in the denominator tends to zero: the forced oscillations increase in amplitude. Consequently a small input amplitude a leads to a much larger output

  amplitude $\dfrac{a\omega^2}{\omega^2 - \Omega^2}$. This effect is known as *resonance*.

- This solution is only valid in cases where Ω does not actually equal ω.

Over the last few pages we have set up and solved a differential equation to model a simple case of forced oscillations. This has given us a mathematical explanation for the phenomenon of resonance which we described at the start of this chapter.

The case when $\Omega = \omega$

Resonance occurs when the frequency of the driving function is the same as the natural frequency of the system. Looking back at the differential equation for the general spring-mass system (equation ② on p116), this occurs when $\Omega = \omega$ and so the differential equation becomes

$$\frac{d^2x}{dt^2} + \omega^2 x = a\omega^2 \sin \omega t.$$

Since it is unaffected by the function on the right hand side of the equation, the complementary function is still $A \sin \omega t + B \cos \omega t$. To obtain the particular integral, given the function on the right side, you would normally try $x = p\sin \omega t + q\cos \omega t$ but this is included in the complementary function. So you multiply the usual trial function by the independent variable.

In this case try $x = t(p\sin \omega t + q\cos \omega t)$.

Differentiating this gives

$$\frac{dx}{dt} = (p\sin\omega t + q\cos\omega t) + t(p\omega \cos\omega t - q\omega \sin\omega t)$$

and

$$\frac{d^2x}{dt^2} = p\omega \cos\omega t - q\omega \sin\omega t + p\omega \cos\omega t - q\omega \sin\omega t + t(-p\omega^2\sin\omega t - q\omega^2\cos\omega t)$$

$$= 2p\omega \cos\omega t - 2q\omega \sin\omega t - p\omega^2 t\sin\omega t - q\omega^2 t\cos\omega t.$$

Substituting these into the differential equation gives

$$2p\omega\cos\omega t - 2q\omega\sin\omega t - p\omega^2 t\sin\omega t - q\omega^2 t\cos\omega t$$
$$+ p\omega^2 t\sin\omega t + q\omega^2 t\cos\omega t \equiv a\omega^2\sin\omega t.$$

Equating coefficients:

$\sin\omega t$:

$$-2q\omega - p\omega^2 t + p\omega^2 t = a\omega^2$$

$$\Rightarrow \qquad q = -\frac{1}{2}a\omega.$$

$\cos\omega t$:

$$2p\omega - q\omega^2 t + q\omega^2 t = 0$$

$$\Rightarrow \qquad p = 0.$$

So a particular integral is $-\dfrac{a\omega t}{2}\cos\omega t$ and the general solution in the case when $\Omega = \omega$ is given by

This is called the forcing term

$$x = A\sin\omega t + B\cos\omega t - \frac{a\omega t}{2}\cos\omega t.$$

Note that as t increases the forcing term dominates the solution. It represents an oscillation whose amplitude is proportional to t and so grows linearly with time. This is a mathematical description of resonance. It occurs when the forcing frequency is identical to the natural frequency of the system.

Drawing graphs of forced oscillations

To illustrate the general results we have just established, we take a particular set of values for the variables involved and specify the initial conditions.

The general solution of $\dfrac{d^2x}{dt^2} + \omega^2 x = a\omega^2\sin\Omega t$ for $\Omega \neq \omega$ is given by

$$x = A\sin\omega t + B\cos\omega t + \frac{a\omega^2}{\omega^2 - \Omega^2}\sin\Omega t$$

with $\qquad \omega = \sqrt{\dfrac{k}{m}}.$

We take the following values:

the stiffness of the spring: $\qquad k = 20\ \text{Nm}^{-1}$
the mass of the object: $\qquad m = 0.2\ \text{kg}$
the amplitude of the forcing motion: $\quad a = 0.02\ \text{m}$
acceleration due to gravity: $\qquad g = 10\ \text{ms}^{-2}$,

and we assume that initially the object is stationary at the equilibrium position so that when $t = 0$, $x = 0$ and $\dfrac{dx}{dt} = 0.$

Activity

Show that under these circumstances the object's displacement, x, is given by

$$x = \frac{1}{(100 - \Omega^2)} (2 \sin \Omega t - 0.2\Omega \sin 10t) \qquad (\Omega \neq 10).$$

The graphs in figure 8.3 show the variation of x with t for various values of Ω.

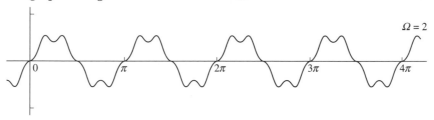

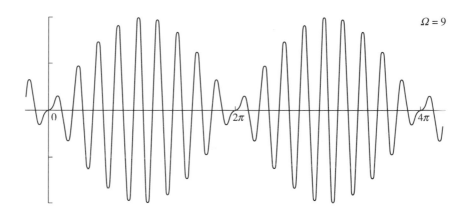

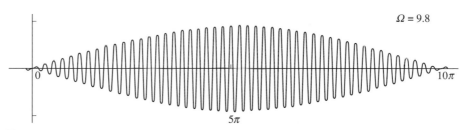

Figure 8.3

Activity

In the case when $\Omega = \omega$, the general solution was found to be

$$x = A \sin \omega t + B \cos \omega t - \frac{a\omega t}{2} \cos \omega t.$$

Show that with the values given on p118, this result may be written as

$$x = 0.01 \sin 10t - 0.1t \cos 10t$$

Figure 8.4 shows the graph of this solution.

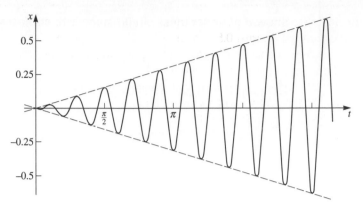

Figure 8.4

The experiment which follows will allow you to see how the theory we have just developed matches what happens in practice.

Experiment

Use apparatus as shown in the diagram to investigate the effects of forced oscillations on a spring-mass oscillator.

One end of the spring is attached to an object, and the other end is attached to an inelastic string passing over a pulley and tied to a peg on a rotating cam. You need to be able to run the motor at different speeds so that the cam rotates at a range of angular speeds.

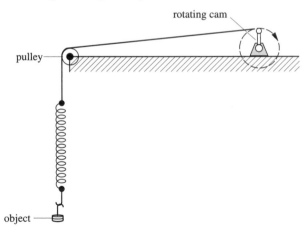

As the cam rotates with a particular (constant) angular speed Ω, the top of the spring is forced to vibrate so that its displacement is $a \sin \Omega t$ where a and Ω are constants. Design and carry out an experiment to investigate the forced oscillations of the system with a range of forcing frequencies, and to demonstrate the occurrence of resonance.

Repeat the experiment with objects of different mass, and validate the result that the resonant frequency is proportional to $\sqrt{\dfrac{1}{m}}$.

1. A spring-mass system consists of a spring of natural length 0.5 m and stiffness 30 Nm⁻¹ and an object of mass 0.6 kg. The system is placed horizontally on a smooth table, with one end attached to a point which is forced to vibrate so that its displacement is $y = 0.1 \cos \Omega t$.

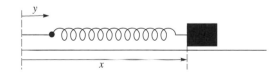

(i) Show that the position x of the object satisfies the differential equation

$$\frac{d^2x}{dt^2} + 50x = 25 + 5\cos \Omega t$$

(ii) If the object is initially at rest in its equilibrium position, find the particular solution for x, where $\Omega \neq \sqrt{50}$.

(iii) Use a graphics calculator or computer to draw the graph of the function x for different values of Ω. Describe the motion of the object.

(iv) State the value of Ω for which resonance occurs. Find the particular solution for x in this case, and sketch its graph.

2. An electrical circuit consists of a 1 henry inductor and a 10^{-4} F capacitor in series with a sinusoidal power source.

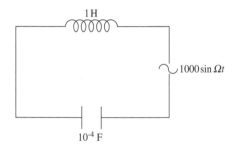

The charge q, in coulombs, stored in the capacitor is given by the differential equation

$$\frac{d^2q}{dt^2} + 10000q = 1000\sin\Omega t$$

where t is the time in seconds. Initially the charge q and current $\frac{dq}{dt}$ in the circuit are both zero.

(i) Find the particular solution for the charge q.

(ii) State the value of Ω for which resonance occurs. Find the particular solution for $q(t)$ in this case. Calculate the time at which the charge first exceeds 10 000 coulombs.

(iii) Use a graphics calculator or computer to graph the solution in (i), for different values of Ω, and in (ii).

3. An oscillatory system with negligible damping is forced to vibrate by a periodic force of constant amplitude Γ and frequency $\frac{p}{2\pi}$. The displacement y of the system from its equilibrium position satisfies the differential equation

$$\frac{d^2y}{dt^2} + \omega^2 y = \omega^2 F \sin pt.$$

(i) Find the general solution of this equation.

(ii) Obtain the particular solution for which $y = 1$ and $\frac{dy}{dt} = 0$ when $t = 0$.

(iii) Use a graphics calculator or computer to draw the graph of the particular solution for $F = 1$ and $p = \pi$ in the cases when (a) $\omega = 0.5\pi$ and (b) $\omega = 0.9\pi$. Describe what happens as ω approaches π.

(iv) Show that if $\omega \approx p$ then, to a good approximation, the particular solution becomes

$$y = \cos \omega t + \frac{F}{2}\sin \omega t - \frac{\omega Ft}{2}\cos \omega t.$$

Damped forced oscillations

In Chapter 7 you saw the effect on simple harmonic motion of introducing a linear damping device called a dashpot. The oscillations, if they occurred, decayed to zero. If the damping constant r was large compared with the stiffness of the spring and the mass of the object, oscillations did not occur at all.

Earlier in this chapter you have seen the effect of forcing an undamped system. Most real systems do have an element of damping, so in this section we explore the effect of including a linear dashpot in the system. To make comparison easy, we shall look again at the spring-mass oscillator with the values given on p118.

Figure 8.5 shows the spring-mass system, which as before is forced to oscillate. A linear dashpot has been added below the object.

Figure 8.5

As before, the spring has stiffness $20\,\text{Nm}^{-1}$, the object has mass $0.2\,\text{kg}$ and the amplitude of the forcing oscillation is $2\,\text{cm}$. The dashpot constant is $1\,\text{Nm}^{-1}\text{s}$ and g is taken to be $10\,\text{ms}^{-2}$. The object starts from rest in its equilibrium position.

The first step is to formulate a differential equation to model this system.

Figure 8.6 shows the system and the forces acting, first in equilibrium and then at some general time t during the motion.

In equilibrium, the extension of the spring is $e = \dfrac{mg}{k} = \dfrac{2}{20} = 0.1$. At the general time t the object is displaced a distance x below the equilibrium level.

The acceleration of the object is $\dfrac{d^2x}{dt^2}$.

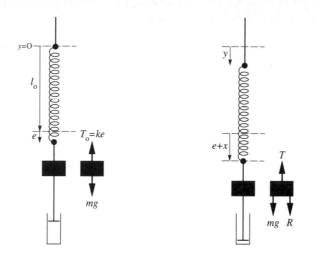

Figure 8.6

The net force on the object in the direction of positive x (i.e. downwards) is $mg + R - T$.

Applying Newton's Second Law at any instant t gives

$$m\frac{d^2x}{dt^2} = mg + R - T.$$

The extension of the spring is $e + x - y$ so $T = 20(0.1 + x - y)$.

Let the length of the dashpot at time t be L, and its length in equilibrium be L_0. Then $L = L_0 - x$, and $\dfrac{dL}{dt} = -\dfrac{dx}{dt}$.

The dashpot force R is given by $r\dfrac{dL}{dt}$, where r is the dashpot constant. So in this case, $R = 1\dfrac{dL}{dt} = -\dfrac{dx}{dt}$.

The equation of motion becomes ⟵ ⟨This is mg⟩

$$0.2\frac{d^2x}{dt^2} = 2 - \frac{dx}{dt} - 20(0.1 + x - y)$$

$$\Rightarrow \qquad \frac{d^2x}{dt^2} + 5\frac{dx}{dt} + 100x = 100y.$$

If the displacement of the forcing point is $y = 0.02 \sin \Omega t$, the differential equation modelling the system is

$$\frac{d^2x}{dt^2} + 5\frac{dx}{dt} + 100x = 2\sin \Omega t.$$

The next stage is to solve the differential equation, which is (as before) a non-homogeneous linear equation with constant coefficients. Its auxiliary equation is

$$\lambda^2 + 5\lambda + 100 = 0$$

whose roots are

$$\lambda = \frac{-5 \pm \surd(375)j}{2}$$

$$\Rightarrow \qquad \lambda = -2.5 \pm 9.68j.$$

The complementary function is therefore $e^{-2.5t}(A\sin 9.68t + B\cos 9.68t)$.

For the particular integral, try $x = p\sin \Omega t + q\cos \Omega t$.

Differentiating this gives

$$\frac{dx}{dt} = p\Omega\cos\Omega t - q\Omega\sin\Omega t$$

and

$$\frac{d^2x}{dt^2} = -p\Omega^2\sin\Omega t - q\Omega^2\cos\Omega t.$$

Substituting these in the differential equation for the system, gives

$$-p\Omega^2\sin\Omega t - q\Omega^2\cos\Omega t + 5p\Omega\cos\Omega t - 5q\Omega\sin\Omega t + 100p\sin\Omega t +$$

$$100q\cos\Omega t \equiv 2\sin\Omega t.$$

Equating coefficients:

$\sin\Omega t$: $\qquad\qquad\qquad -p\Omega^2 - 5q\Omega + 100p = 2.$

$\cos\Omega t$: $\qquad\qquad\qquad -q\Omega^2 + 5p\Omega + 100q = 0.$

Solving these equations for p and q gives

$$p = \frac{2(100 - \Omega^2)}{(100 - \Omega^2)^2 + 25\Omega^2}$$

and

$$q = -\frac{10\Omega}{(100 - \Omega^2)^2 + 25\Omega^2}.$$

The particular integral is therefore

$$\frac{2(100 - \Omega^2)}{(100 - \Omega^2)^2 + 25\Omega^2}\sin\Omega t - \frac{10\Omega}{(100 - \Omega^2)^2 + 25\Omega^2}\cos\Omega t.$$

The general solution is the sum of the complementary function and the particular integral:

$$x = e^{-2.5t}(A\sin 9.68t + B\cos 9.68t) +$$

$$\frac{2(100 - \Omega^2)}{(100 - \Omega^2)^2 + 25\Omega^2}\sin\Omega t - \frac{10\Omega}{(100 - \Omega^2)^2 + 25\Omega^2}\cos\Omega t.$$

As t increases, the natural damped oscillations, given by the complementary function, decay because of the $e^{-2.5t}$ term, leaving

$$x = \frac{2(100 - \Omega^2)}{(100 - \Omega^2)^2 + 25\Omega^2}\sin\Omega t - \frac{10\Omega}{(100 - \Omega^2)^2 + 25\Omega^2}\cos\Omega t.$$

This is called the *steady state* solution. It is the particular integral of the differential equation. It describes the oscillations that occur after the unforced oscillations have died away. Figure 8.7 shows graphs of this steady state solution for two values of Ω.

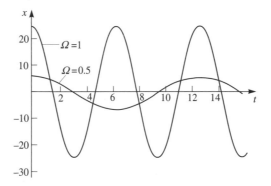

Figure 8.7

Remember that in the undamped case the value of Ω for resonance was calculated by setting the denominator to zero in the particular integral. Catastrophic resonance does not occur in the damped case, because the denominator of each part of the particular integral is always greater than zero. However, the amplitude of the forced vibrations does still depend on the value of Ω.

The amplitude of the steady state oscillations is the square root of the sum of the squares of the coefficients of $\cos \Omega t$ and $\sin \Omega t$ in the steady state solution, i.e.

$$\sqrt{\left[\left(\frac{2(100-\Omega^2)}{(100-\Omega^2)^2+25\Omega^2}\right)^2+\left(\frac{-10\Omega}{(100-\Omega^2)^2+25\Omega^2}\right)^2\right]}.$$

This can be simplified to $\dfrac{2}{\sqrt{(100-\Omega^2)^2+25\Omega^2}}$

This result shows how the amplitude of the steady state solution depends on Ω. Figure 8.8 shows a graph of this amplitude against Ω.

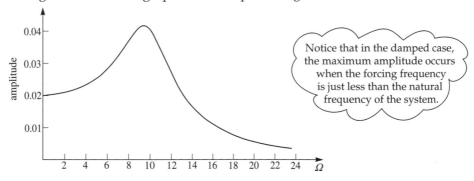

Notice that in the damped case, the maximum amplitude occurs when the forcing frequency is just less than the natural frequency of the system.

Figure 8.8

When Ω and ω are close in value to each other, then it follows that the forcing frequency $\dfrac{\Omega}{2\pi}$ and the natural frequency $\dfrac{\omega}{2\pi}$ are close in value. When this is the case, the steady state oscillations become large compared with the input amplitude a. The motion of the system at these relatively large amplitudes is still called resonance, though the amplitude of the vibrations does not increase without limit as it does in the undamped case.

You have now seen the effect of linear damping on the system. What would be the effect of varying the damping constant, r? To predict this, look at the differential equation for the same damped spring-mass system ($m = 0.2$ and $k = 20$), but this time use a general damping constant r. The equation becomes

$$0.2\frac{d^2x}{dt^2} + r\frac{dx}{dt} + 20x = 0.4\sin\Omega t.$$

We know that the complementary function decays and that the steady state oscillations are given by the particular integral. In this case it is

$$x = \frac{0.4(20 - 0.2\Omega^2)}{(20 - 0.2\Omega^2)^2 + r^2\Omega^2}\sin\Omega t - \frac{0.4r\Omega}{(20 - 0.2\Omega^2)^2 + r^2\Omega^2}\cos\Omega t.$$

The amplitude of the steady state forced oscillation is

$$\frac{2}{\sqrt{(100 - \Omega^2)^2 + 25r^2\Omega^2}}.$$

Figure 8.9 shows graphs of the steady state amplitude against Ω for different values of r.

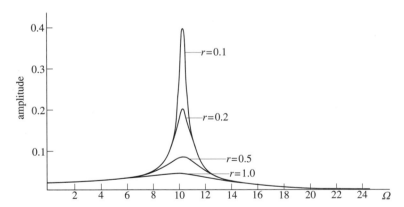

Figure 8.9

The graphs show that as r decreases (i.e. the amount of damping is reduced), the amplitude at the resonant frequency increases. In each case resonance occurs when Ω is very near in value to ω (in this case 10). In any real system there is always some damping but, as you can see, if the damping constant is small the resonance can still be damaging.

Activity

The previous example involved a particular case of damped forced motion in which the various parameters of the system were given particular values.

(i) Show that the differential equation modelling the general case may be written

$$\frac{d^2x}{dt^2} + \alpha\frac{dx}{dt} + \omega^2 x = a\omega^2 \sin \Omega t$$

(ii) Find (a) the general solution; (b) the particular solution corresponding to $x = 0$ and $\dfrac{dx}{dt} = 0$ at $t = 0$; (c) the amplitude of the steady state oscillations.

You can confirm your answers by looking at the Key Points at the end of this chapter.

Exercise 8B

1. The diagram shows a damped spring-mass system consisting of a particle of mass $1\,\text{kg}$ oscillating on a smooth horizontal air track connected by a spring of stiffness $10\,\text{Nm}^{-1}$, natural length 1 metre, to a fixed point O. The particle is attached to a forcing point A via a dashpot of constant $3\,\text{Nm}^{-1}\text{s}$. The displacement of point A from the fixed point O is $y\,(t)$.

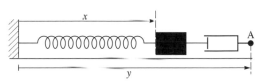

(i) Show that the displacement x of the particle satisfies the differential equation

$$\frac{d^2x}{dt^2} + 3\frac{dx}{dt} + 10x = 10 + 3\frac{dy}{dt}.$$

(ii) If the forcing function is $y = 2 + 0.1\cos 2t$, and the particle starts from rest at $x = 1$, find the particular solution of the differential equation.

(iii) Use a graphics calculator or computer to draw graphs of x and y against t on the same axes. Describe the motion of the particle and compare its motion with that of the point A.

2. The current I amperes in an electric circuit (consisting of inductance and resistance in series with a sinusoidal power source) is given by the differential equation

$$\frac{dI}{dt} + 2I = \sin 3t,$$

where t is the elapsed time in seconds. Find

(i) the complementary function;

(ii) the particular integral;

(iii) the current I as a function of time t, given that initially $I = 0$;

(iv) the amplitude of the oscillations of the current after a long time has elapsed.

[MEI]

3. The forced oscillations of a damped harmonic system are modelled by the differential equation

$$\frac{d^2x}{dt^2} + 2\frac{dx}{dt} + 10x = \sin \omega t.$$

(i) Find the complementary function of this differential equation.

(ii) Hence determine the nature of the damping.

(iii) Find the particular integral.

(iv) Show that the amplitude of the oscillations of the system when t is large is given approximately by

$$\frac{1}{\sqrt{\{(10 - \omega^2)^2 + 4\omega^2\}}}.$$

[MEI]

DE

Exercise 8B continued

4. An electrical circuit consists of a 100 ohm resistor, a 1 henry inductor and a 10^{-4} farad capacitor in series with a sinusoidal power source. The charge q coulombs stored in the capacitor at time t is given by the differential equation

$$\frac{d^2q}{dt^2} + 100\frac{dq}{dt} + 10\,000q = 1000\sin 100t$$

where t is the time in seconds.

(i) Find the complementary function of the differential equation.

(ii) Decide if the circuit is overdamped, critically damped or underdamped.

(iii) Find the particular integral and hence the general solution.

(iv) Initially both q and $\dfrac{dq}{dt}$, the current in the circuit, are zero. Find the values of the unknown constants in the general solution.

(v) Use a graphics calculator or computer to sketch the solution and describe how the charge in the circuit changes with time.

5. (i) Find the general solution of the simple harmonic differential equation
$$\ddot{x} + \omega^2 x = 0.$$

(ii) Comparing each of the following differential equations with the one above, give in each case a physical interpretation of the extra term.

$$\ddot{x} + k\dot{x} + \omega^2 x = 0, \qquad k > 0,$$
$$\ddot{x} \qquad + \omega^2 x = A\cos \alpha t$$

(iii) Solve the second differential equation in part (ii) for $\alpha \neq \omega$.

(iv) Explain what happens to your solution as α approaches the value of ω.

[MEI]

6. A sphere of mass m and radius a is falling vertically through a liquid which produces a linear resistance force. The motion of the sphere is modelled by the differential equation

$$m\frac{d^2x}{dt^2} + ra\frac{dx}{dt} = mg$$

where x is the distance fallen in t seconds and r is a constant. The sphere is released from rest so that $x = 0$ and $\dfrac{dx}{dt} = 0$ when $t = 0$.

(i) Find the solution for x and hence the velocity of the sphere as a function of time.

(ii) Draw a graph of the velocity against time, and describe the motion of the sphere.

7. You are to model the vertical vibrations of a car driving along a bumpy road. The diagram shows the car, modelled by a particle of mass m, and the suspension system, modelled by a spring of stiffness k and a linear damping device of damping constant r. The bumps in the road are modelled by taking the road surface to be a sinusoidal curve with period d metres and amplitude h metres.

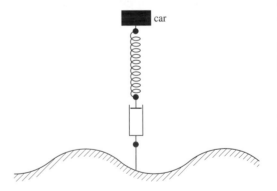

At a general time t, let the height of the particle above the road surface be x and the height of the road surface above its average level be y.

(i) Show that the equation of the particle modelling the car is

$$m\frac{d^2x}{dt^2} + r\frac{dx}{dt} + kx = kl - mg - m\frac{d^2y}{dt^2}$$

where l is the natural length of the spring.

(ii) Show that, when the car is driven at speed v, the function y is given by

$$y = h\sin\left(\frac{2\pi v}{d}t\right).$$

(iii) The following data refer to a car being driven over an unmade road such as might be found in the Australian outback.

$m = 800\,\text{kg}, \quad k = 8 \times 10^{-5}\,\text{Nm}^{-1},$
$r = 9 \times 10^{3}\,\text{Nm}^{-1}\text{s}, \quad d = 2\,\text{m}$
$h = 5$ cm and $l = 0.4$ m.

Find an expression for the amplitude of the steady state oscillatory solution of the equation of motion in terms of the speed v, of the car (in ms^{-1}).

(iv) Draw a graph of the amplitude against v. What advice would you give the driver?

(v) What difference would it have made to your answer in part (iv) if the bumps on the road were closer together?

8. During the design of a car, part of the suspension system is tested by subjecting it to violent displacements. One such test is modelled by the differential equation

$$\ddot{x} + 2k\dot{x} + x = 1$$

where x is displacement, and initially $x = 0$ and $\dot{x} = 0$. The parameter k (> 0) is known as the damping coefficient and can be varied during the tests.

For optimum road holding a "hard" suspension is desirable and it is believed that to achieve this the damping should be critical.

(i) Find the value of k for critical damping.

(ii) Determine x as a function of time t in this case.

For a more comfortable ride a "soft" suspension is proposed in which $k = 0.6$.

(iii) Determine x as a function of time t for the "soft" suspension.

(iv) Find the maximum displacement of the "soft" suspension.

[MEI]

Investigation

Oscillations of a mass-pulley system

The diagram shows a simple mass–pulley system.

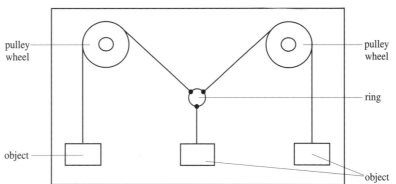

Investigate the motion of the three objects when the central object is displaced from its equilibrium position.

Can the motion of the central object be described as damped simple harmonic motion?

Investigations continued

Electric circuit

Set up a circuit in which an inductor, a capacitor and a resistor are in series with an applied voltage. Investigate and analyse the behaviour when the voltage is provided by

(i)   a sequence generator
(ii) a variable frequency oscillator.

KEY POINTS

- Motion for which the differential equation is $\dfrac{d^2x}{dt^2} + \omega^2 x = \omega^2 f(t)$ is

  called undamped forced harmonic motion; $f(t)$ is the forcing term. For a sinusoidal forcing term, $f(t) = a \sin \Omega t$ and the general solution of the differential equation is

$$x = A \sin \omega t + B \cos \omega t + \frac{a\omega^2}{\omega^2 - \Omega^2} \sin \Omega t \qquad (\omega \neq \Omega).$$

- The quantity $\dfrac{\Omega}{2\pi}$ is called the forcing frequency, and $\dfrac{\omega}{2\pi}$ is the

  natural frequency of the system. When $\Omega = \omega$ resonance occurs. In

  this case $f(t) = a \sin \omega t$ and the general solution is

$$x = A \sin \omega t + B \cos \omega t - \frac{a\omega t}{2} \cos \omega t.$$

- With linear damping, the differential equation model becomes

$$\frac{d^2x}{dt^2} + \alpha \frac{dx}{dt} + \omega^2 x = a\omega^2 \sin \Omega t \qquad (\alpha > 0).$$

- The complementary function of this differential equation decays to zero. The particular integral remains and is called the *steady state* solution. If $f(t) = a \sin \Omega t$ then the steady state solution is

$$x = \frac{(\omega^2 - \Omega^2)a\omega^2}{(\omega^2 - \Omega^2)^2 + \Omega^2\alpha^2} \sin \Omega t - \frac{a\alpha\Omega\omega^2}{(\omega^2 - \Omega^2)^2 + \Omega^2\alpha^2} \cos \Omega t$$

- The amplitude of this steady state solution is $\dfrac{a\omega^2}{\sqrt{(\omega^2 - \Omega^2)^2 + \Omega^2\alpha^2}}$.

- When Ω takes the value which maximises the amplitude of the steady

  state oscillations, $\dfrac{\Omega}{2\pi}$ is called the *resonant frequency*.

9 Systems of differential equations

What's lost upon the roundabout
We pulls up on the swings!

Patrick Reginald Chambers

The photograph shows a typical Los Angeles Smog, caused largely by exhaust fumes from motor vehicles. This smog, like almost all forms of man-made pollution, constitutes a health hazard to those who are exposed to it. The routes by which noxious pollutants enter human beings often involve several stages, for example a food chain.

Consequently when formulating differential equations to model such processes, you are likely to require a number of variables, a *system*. This is illustrated by the following example, which is based on studies into the effects of lead pollution on people living and working in Los Angeles. You should however realise from the outset that systems of differential equations may be applied to many different situations; modelling pollution is just one example.

Lead enters the body via food, air and water. It builds up in the blood, in body tissues and in the bones. Lead can leave the body via urine, hair, nails, skin and sweat, but enough can remain to present a significant health risk. It takes longer for lead to get into the bones than into the blood and the body tissue, but once there it is harder to remove and is particularly harmful. We shall concentrate first on the lead building up in the blood and body tissue.

If we represent the blood and body tissue as two separate compartments in a diagram, and show by arrows the possible transfers of lead into and out of each compartment, we can build up what is called a *compartmental model* (figure 9.1).

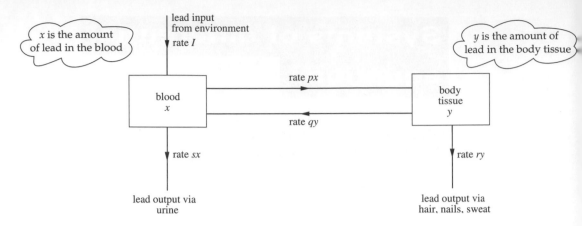

Figure 9.1

Biological research shows that it is reasonable to assume that the rate of transfer of lead from one part of the body to another is proportional to the amount of lead present. The symbols p, q, r and s are the constants of proportionality for the routes shown in the diagram.

A mathematical model can be built up for this system. Take the blood first, at any time t:

rate of increase of lead in bloodstream = rate of input − rate of output.

Expressing this mathematically using the notation in the flowchart,

$$\frac{dx}{dt} = (I + qy) - (px + sx)$$
$$= I - (p + s)x + qy.$$

Similarly for the body tissue

$$\frac{dy}{dt} = px - qy - ry$$
$$= px - (q + r)y.$$

We now have two differential equations, both involving the dependent variables x and y and the independent variable t. When two or more differential equations involve the same combination of variables they are called a *system* of differential equations. In this case, since both of the equations are linear, we have a linear system. In this chapter you will meet methods of solving linear systems of differential equations analytically.

Solving linear simultaneous differential equations

Before taking the example of lead pollution further, we take a simple numerical example to show how linear first order differential equations may be solved.

To see the method of approach, look at the pair of equations

$$\frac{dx}{dt} = 2x + 4y \qquad \qquad ①$$

$$\frac{dy}{dt} = x - y \qquad \qquad ②$$

with initial conditions $x = 2$ and $y = -2$ when $t = 0$.

We start by making x the subject in equation ② (since this is the simpler equation):

$$x = \frac{dy}{dt} + y. \qquad \qquad ③$$

Now we differentiate both sides with respect to t to obtain

$$\Rightarrow \quad \frac{dx}{dt} = \frac{d^2y}{dt^2} + \frac{dy}{dt}.$$

Substituting for x and $\frac{dx}{dt}$ in equation ① gives

$$\frac{d^2y}{dt^2} + \frac{dy}{dt} = 2\left(\frac{dy}{dt} + y\right) + 4y$$

This simplifies to the second order differential equation with constant coefficients:

$$\frac{d^2y}{dt^2} - \frac{dy}{dt} - 6y = 0. \qquad \qquad ④$$

This equation involves only one of the dependent variables, in this case y but not x. We can apply standard techniques to solve this.

The auxiliary equation is $\lambda^2 - \lambda - 6 = 0$ which has roots $\lambda = 3$ and $\lambda = -2$.

The general solution of equation ④ is therefore

$$y = Ae^{3t} + Be^{-2t}.$$

Substituting this into equation ③ gives the general solution for x:

$$x = 4Ae^{3t} - Be^{-2t}.$$

We now have the general solution, since we have an expression for x in terms of t and a separate expression for y in terms of t.

To find the values of the unknown constants of integration A and B we use the given initial conditions:

$t = 0,\ x = 2 \quad \Rightarrow \quad 2 = 4A - B$
$t = 0,\ y = -2 \quad \Rightarrow \quad -2 = A + B$

Solving for A and B gives $A = 0$ and $B = -2$.

The particular solution of the system of equations ① and ② satisfying the given initial conditions is

$$x = 2e^{-2t}$$
$$y = -2e^{-2t}.$$

Both x and y tend to zero as t tends to infinity. Figure 9.2 shows the behaviour of the solution graphically.

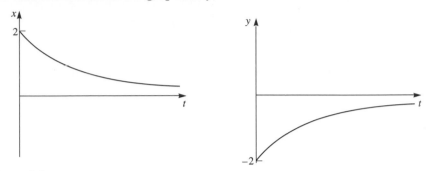

Figure 9.2

Sometimes it is important to understand the relationship between x and y, in which case the solution can be thought of as a pair of parametric equations with parameter t. In the example above t can be eliminated simply by adding the two solutions:

$$x = 2e^{-t}$$
$$y = -2e^{-t}$$
$$\overline{x + y = 0.}$$

This is illustrated in figure 9.3. Notice that only part of the line $x + y = 0$ is required, since the starting point is $x = 2$, $y = -2$ and as $t \to \infty$, $(x, y) \to (0, 0)$.

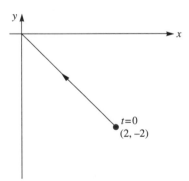

Figure 9.3

A graph like this is called a *solution curve*. It shows the relationship between the dependent variables (in this case x and y) as the independent variable (in this case t) increases through its permitted range.

EXAMPLE

(i) Find the general solution of the system of differential equations

$$\frac{dx}{dt} = -3y \qquad \qquad ①$$

$$\frac{dy}{dt} = 3x \qquad \qquad ②$$

(ii)   Find the particular solution given that $x = 3$, $y = 4$ when $t = 0$.

(iii) Sketch graphs of x and y against time t, and of y against x (the solution curve) for this particular solution. Describe the behaviour of the system.

Solution

(i)   From equation ②

$$x = \frac{1}{3}\frac{dy}{dt}.$$

Differentiating this gives

$$\frac{dx}{dt} = \frac{1}{3}\frac{d^2y}{dt^2}.$$

Substituting for $\dfrac{dx}{dt}$ in equation ① gives a second order differential equation with constant coefficients:

$$\frac{d^2y}{dt^2} = -9y.$$

You will recognise this as the equation of simple harmonic motion with general solution

$$y = A \sin 3t + B \cos 3t.$$

Substituting for y in equation ② gives the general solution for x:

$$x = A \cos 3t - B \sin 3t .$$

(ii)   Applying the given boundary conditions
$x = 3, t = 0 \;\Rightarrow\; 3 = A$
$y = 4, t = 0 \;\Rightarrow\; 4 = B$
The particular solution is

$$x = 3 \cos 3t - 4 \sin 3t$$
$$y = 3 \sin 3t + 4 \cos 3t$$

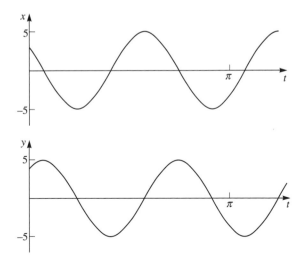

(iii) These solutions show that x and y are periodic functions of time with the same period, $\dfrac{2\pi}{3}$. Both x and y have amplitude $\sqrt{(3^2 + 4^2)} = 5$, but they are out of phase. The solution of the curve is a circle of radius 5.

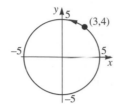

A pendulum that oscillates in both the x and the y directions could be modelled by this pair of differential equations.

EXAMPLE

A system of differential equations is given by

$$\frac{dx}{dt} = -x + y - 1 \qquad \qquad \text{①}$$

$$\frac{dy}{dt} = -x - y + 3. \qquad \qquad \text{②}$$

When $t = 0$, $x = 0$ and $y = 3$.
(i) Find expressions for x and y in terms of t.
(ii) Draw the graph of y against x for values of $t \geq 0$. Describe what happens as $t \to \infty$.

Solution

(i) Equation ① gives

$$y = x + \frac{dx}{dt} + 1. \qquad \qquad \text{③}$$

Differentiating this to find an expression for $\dfrac{dy}{dt}$:

$$\frac{dy}{dt} = \frac{dx}{dt} + \frac{d^2x}{dt^2}.$$

Substituting for y and $\dfrac{dy}{dt}$ in equation ② gives

$$\frac{dx}{dt} + \frac{d^2x}{dt^2} = -x - \left(x + \frac{dx}{dt} + 1\right) + 3$$

Rearranging this, we find that the pair of first order differential equations gives rise to a second order differential equation with constant coefficients:

$$\frac{d^2x}{dt^2} + 2\frac{dx}{dt} + 2x = 2.$$

As you may verify, the general solution of this equation is

$$x = Ae^{-t} \sin t + Be^{-t} \cos t + 1.$$

Substituting for x and $\dfrac{dx}{dt}$ in equation ③ gives the general solution for y:

$$y = Ae^{-t} \cos t - Be^{-t} \sin t + 2.$$

The initial conditions give the values of A and B:

when $t = 0, x = 0 \;\Rightarrow\; 0 = B + 1 \;\Rightarrow\; B = -1;$
when $t = 0, y = 3 \;\Rightarrow\; 3 = A + 2 \;\Rightarrow\; A = 1.$

The particular solution that satisfies the initial conditions is therefore

$$x = e^{-t} \sin t - e^{-t} \cos t + 1.$$
$$y = e^{-t} \cos t + e^{-t} \sin t + 2.$$

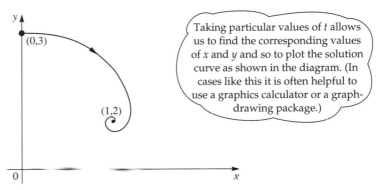

Taking particular values of t allows us to find the corresponding values of x and y and so to plot the solution curve as shown in the diagram. (In cases like this it is often helpful to use a graphics calculator or a graph-drawing package.)

(ii) We see that as t increases, $e^{-t} \to 0$, so $x \to 1$ and $y \to 2$. In the long term the system approaches the point (1, 2).

Solving the lead absorption equations

You now have the techniques to solve the system of equations on page 132 which were developed to model lead absorption and clearance by the human body. The equations are

$$\frac{dx}{dt} = I - (p + s)x + qy$$
$$\frac{dy}{dt} = px - (q + r)y.$$

In a heavy smog in Los Angeles, the intake of lead can be 50 µg per day, so $I = 50$. Data for an average-sized teenager give the following values: $p = 0.01$, $q = 0.01$, $r = 0.02$ and $s = 0.02$, where the masses are in micrograms (µg) and time is in days. Substituting these into the differential equations gives

$$\frac{dx}{dt} = 50 - 0.03x + 0.01y \qquad\qquad ①$$

$$\frac{dy}{dt} = 0.01x - 0.03y. \qquad ②$$

By eliminating y between the two equations you find that x satisfies the second order differential equation

$$100\frac{d^2x}{dt^2} + 6\frac{dx}{dt} + 0.08x = 150$$

The general solution of this equation is

$$x = Ae^{-0.04t} + Be^{-0.02t} + 1875.$$

Substituting for x into ① gives

$$y = -Ae^{-0.04t} + Be^{-0.02t} + 625.$$

Figure 9.4 shows graphs of x and y against t for someone arriving in Los Angeles from a totally unpolluted area. The limiting values of x and y are 1875 and 625. What this means is that a constant input of 50 µg per day will eventually lead to a steady level of lead in the blood of 1875 µg and a steady level of lead in the body tissue of 625 µg.

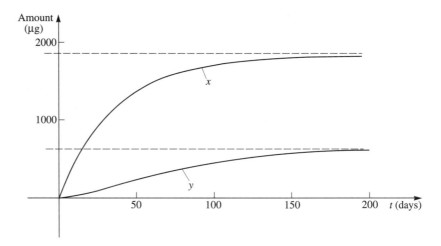

Figure 9.4

Activity

Lead can also be transferred between the blood and the bones. The rate of input to the bones is ux, and the rate of output from the bones is vz, where z is the amount of lead in the bones. Typical values for u and v are 0.004 and 0.000 035. You can assume the amount of lead transferred directly between the bones and the body tissue to be zero.

Extend the model formulated above to include a third compartment, the bones. You will now need three differential equations.

Equilibrium Points

In the example of lead pollution, we found that the levels in the blood and tissue settled on steady equilibrium values of 1875 μg and 625 μg. Many systems settle to such equilibrium (or stationary) points. In the case of lead build-up in the human body, an equilibrium point corresponds physically to the amount of lead remaining constant in both the blood and the body tissue. For this balance to remain, the net exchange between the blood and the body tissue must be zero and the total amount of lead entering the body must be equal to the amount leaving the body.

It is useful to think of the state of a system at any instant being represented by the position of a particle as it travels along the solution curve. Each point on a solution curve has co-ordinates that are the values of x and y corresponding to a particular value of t. The curve starts at the point corresponding to the inital conditions. It finishes at the co-ordinates that correspond either to the limit as $t \to \infty$ or to some other specified end instant. It is important to realise that there is a *direction* associated with a solution curve: the behaviour of the system is described by the path of the curve from the start point to the end point (i.e. in the direction in which time increases). It is usual to indicate this by an arrow.

Equilibrium points can usually best be seen by drawing the family of solution curves for the general solution. The three other examples used so far in this chapter all included a drawing of a particular solution curve; the graphs below show the families of solution curves for their general solutions.

Case 1

$$\frac{dx}{dt} = 2x + 4y$$

$$\frac{dy}{dt} = x - y$$

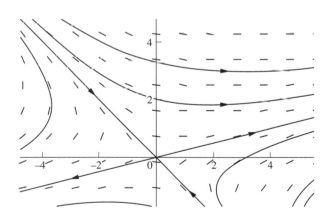

Case 2

$$\frac{dx}{dt} = -3y$$

$$\frac{dy}{dt} = 3x$$

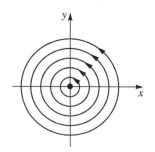

Case 3

$$\frac{dx}{dt} = -x + y - 1$$

$$\frac{dy}{dt} = -x - y + 3$$

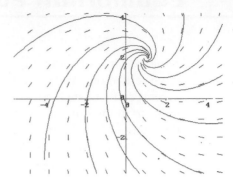

You will see from looking at these graphs that there are differences between the equilibrium points.

- In Case 1 none of the solution curves, apart from the lines $y = -x$ and $y = \frac{1}{4}x$, passes through the equilibrium point O. Further, unless the particle starts on the line $y = -x$ it never reaches O and indeed eventually moves further and further away from it.

- In Case 2 the solution curves circle round the equilibrium point, O, getting neither closer nor further away. The only 'curve' that reaches O is the point itself.

- In Case 3 all the curves converge onto the equilibrium point $(1, 2)$.

Systems, like these three, in which $\dfrac{dx}{dt}$ and $\dfrac{dy}{dt}$ are not dependent on t are called *autonomous*. For autonomous systems, equilibrium points may be found by setting $\dfrac{dx}{dt}$ and $\dfrac{dy}{dt}$ both equal to zero, as in the next example.

EXAMPLE

Find the equilibrium point for the system

$$\frac{dx}{dt} = 3x - 2y$$

$$\frac{dy}{dt} = x - 2y + 4$$

Solution
The equilibrium point is given by

$$3x - 2y = 0$$
$$x - 2y + 4 = 0 \qquad \qquad ②$$

Subtracting ② from ① gives
$$2x - 4 = 0$$
$$\Rightarrow \qquad x = 2$$

Substituting this in either equation gives $y = 3$, so the equilibrium point of the system is $(2, 3)$.

Investigating the solution of a system by looking at its behaviour near an equilibrium point can provide a great deal of qualitative information without requiring detailed mathematical analysis. For a linear system, like the one above, you can solve the equations analytically, then use a parametric plot facility to draw a range of solution curves. For most non-linear systems, though, you cannot solve the equations analytically.

Investigation

Equilibrium points

For a system with two dependent variables (e.g. x and y) there are six types of equilibrium points and these are the subject of the following investigation.

For each of the systems below

(i)   identify the equilibrium point;
(ii) draw the family of solution curves;
(iii) describe the behaviour of the system near the equilibrium point.

(a) $\dfrac{dx}{dt} = 2x + y$

$\dfrac{dy}{dt} = x + 2y$

(b) $\dfrac{dx}{dt} = y$

$\dfrac{dy}{dt} = x$

(c) $\dfrac{dx}{dt} = x - y$

$\dfrac{dy}{dt} = x + y$

(d) $\dfrac{dx}{dt} = -y$

$\dfrac{dy}{dt} = x$

(e) $\dfrac{dx}{dt} = x + y$

$\dfrac{dy}{dt} = y$

(f) $\dfrac{dx}{dt} = x$

$\dfrac{dy}{dt} = y$

Non-linear systems

So far in this chapter we have looked exclusively at linear systems. These have the advantage that the equations can be solved analytically. Unfortunately, though, many real systems are non-linear, and often they cannot be solved analytically. That does not mean that the preceding work is of no use.

For a start, we can use the expressions for $\dfrac{dx}{dt}$ and $\dfrac{dy}{dt}$ to find a single expression for $\dfrac{dy}{dx}$, and plot the tangent field.

EXAMPLE   You are given the non-linear system of differential equations

$$\frac{dx}{dt} = x - y^2$$

$$\frac{dy}{dt} = x^2 - y$$

(i)   Draw the tangent field for the relationship between y and x.

(ii) Draw a number of solution curves.

Solution

(i)   An expression for the gradients of the direction indicators may be found by using

$$\frac{dy}{dx} = \frac{\dfrac{dy}{dt}}{\dfrac{dx}{dt}} = \frac{x^2 - y}{x - y^2}$$

This allows the value of $\dfrac{dy}{dx}$ to be calculated at any point.

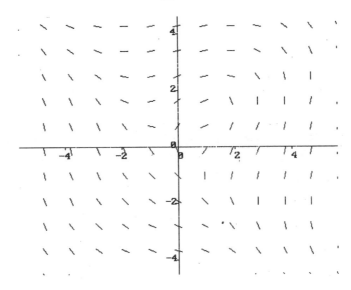

(ii) Several solution curves are shown on the graph below.

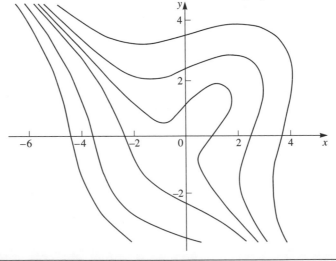

Investigation

These solution curves do not tell you what happens in the first quadrant near the origin. Investigate the behaviour of the curve there.

Further, by investigating equilibrium points and using numerical methods akin to Euler's method (Chapter 5), it is often possible to investigate a non-linear system thoroughly. A spreadsheet package is very helpful in this.

The next example involves a non-linear predator prey model.

EXAMPLE

A closed environment (such as a small island) supports populations of rabbits and foxes. At time t years their sizes are given by r and f. The rabbits feed off the vegetation and the foxes eat the rabbits.

A proposed model for this situation is given by the pair of differential equations

This equation is non-linear because it contains a term in r^2 →

$$\frac{dr}{dt} = k_1 r \left(1 - \frac{r}{r_m} \right) - \lambda f$$

$$\frac{df}{dt} = k_2 f \left(\alpha \frac{r}{f} - 1 \right).$$ ← This equation is linear

The meanings of the various terms in this model are as follows:
r_m is the greatest number of rabbits the environment can support;
α is the ratio of foxes to rabbits above which the number of foxes start to decline due to starvation;

k_1, k_2 are constants related to the number of offspring per rabbit and fox respectively in one year;

λ is the number of rabbits eaten per fox in a year.

Following preliminary investigations, the values below are assigned to these parameters:

$$r_m = 12\,000 \qquad \alpha = 0.01$$
$$k_1 = 3 \qquad\qquad k_2 = 1$$
$$\lambda = 50$$

(i) Substitute the assigned values into the differential equations and find the values of r and f at the equilibrium points.

(ii) Use a step-by-step method with steps of one year to estimate the populations for the next 25 years, given that their initial values are $r = 400$ and $f = 2$.

Draw graphs of f against t, r against t and f against r.

Solution

(i) Substituting the assigned parameter values gives

$$\frac{dr}{dt} = 3r \left(1 - \frac{1}{12\,000} r \right) - 50f$$ ①

$$\frac{df}{dt} = 0.01r - f \qquad ②$$

At an equilibrium point both $\frac{dr}{dt} = 0$ and $\frac{df}{dt} = 0$, so that (in equation ②)

$$0.01r - f = 0$$
$$\Rightarrow \qquad r = 100f.$$

Substituting for r in equation ① with $\frac{dr}{dt} = 0$ gives

$$300f\left(1 - \frac{f}{120}\right) - 50f = 0$$

$$\Rightarrow \qquad f^2 - 100f = 0$$
$$\Rightarrow \qquad f = 0 \text{ or } 100.$$

The corresponding values of r are

$$r = 0 \text{ or } 10\,000.$$

There are two equilibrium points: $r = 0, f = 0$ and $\qquad r = 10\,000, f = 100$.

(iii) Using a spreadsheet gives value of r and f as shown below.

time	rabbits	foxes
0	400	2
1	1460	4
2	5107.1	14.6
3	13177.782	51.071
4	6744.0924	131.77782
5	9016.7828	67.440924
6	12369.492	90.167828
7	6718.4935	123.69492
8	9404.6893	67.184935
9	12147.465	94.046893
10	6997.2884	121.47465
11	9674.9097	69.972884
12	11800.025	96.749097
13	7552.4973	118.00025
14	10049.923	75.524973
15	11173.206	100.49923
16	8457.7303	111.73206
17	10361.018	84.577303
18	10377.533	103.61018
19	9406.3247	103.77533
20	10316.796	94.063247
21	9954.9517	103.16796
22	9886.1429	99.549517
23	10133.14	98.861429
24	9919.3566	101.3314
25	10012.447	99.193566

These results are illustrated on the graphs that follow.

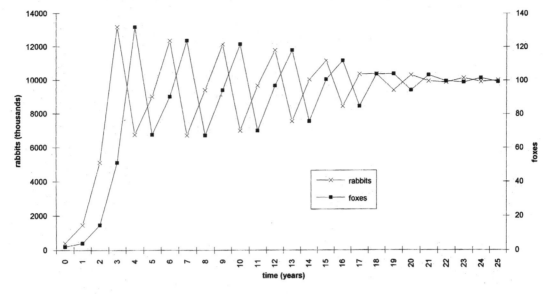

Graph showing the sizes of the rabbit and fox populations with time (notice the different vertical scales.

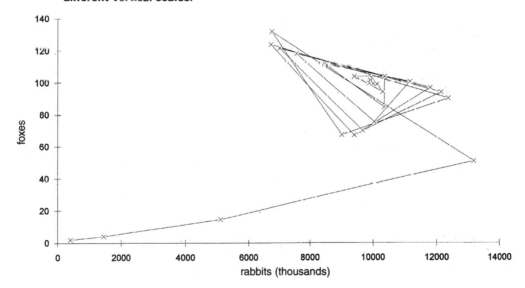

Solution 'curve' showing the variation of the number of foxes with the number of rabbits.

This example illustrates several important points

- Even though you cannot solve the non-linear system analytically you are able to investigate the system using familiar methods.
- You can use step-by-step methods to investigate the behaviour of non-linear systems.
- Using a spreadsheet means you can investigate the system with different initial conditions. It is also easy to see what happens if you change the values of the assigned parameters.

If you try different initial conditions in this example (with the assigned parameters), you will find that one of three patterns emerges:

- convergence to the non-zero equilibrium point as in the case above: this corresponds to stable populations;
- convergence to the zero equilibrium point: the populations die out;
- a three year cycle between certain pairs of values of r and f, none of which is an equilibrium point.

Population cycles occur naturally; the best known is a 10-year cycle exhibited by many Arctic mammals in Canada and Siberia (and also Scotland). Data collected by the Hudson Bay Company on fur sales go back over 250 years and show this cycle for, among other species, the Canadian Lynx and the Arctic Hare. Other animals are found to follow a four year cycle. The model in this example is a simple one involving only two species; real food chains are usually more complicated.

For discussion

It would seem that the accuracy could be improved by taking shorter step lengths. Given the breeding patterns of rabbits and foxes, would doing so improve the solution?

Investigation

Investigate the effect of changing the parameters in the rabbit and fox example.

For discussion

Why is it important to be able to construct mathematical models of animal populations?

The study of non-linear systems of differential equations has only just begun. Until the invention of computers, the standard method of dealing with a non-linear system was to try to transform it into a linear one; if this could not be done, there was little more that could be achieved. The application of computers to non-linear systems has opened up a whole new area of mathematics often known as *Chaos*. Many of their solutions show instability: a small change in the initial conditions produces a large change in the solution, a phenomenon known as the *Butterfly Effect*. However, there are found to be patterns of order within such instability and these can often be represented by beautiful fractal diagrams.

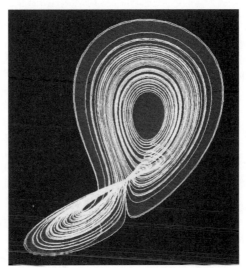

Exercise 9A

1. For each of the following systems of equations:

  (i) solve the equations to find expressions for x and y in terms of t;

  (ii) find the particular solutions for which $x = 1$ and $y = 2$ at $t = 0$;

  (iii) describe the long term behaviour of the system.

(a) $\dfrac{dx}{dt} = 3x - y$

    $\dfrac{dy}{dt} = 2x$

(b) $\dfrac{dx}{dt} = x + y$

    $\dfrac{dy}{dt} = x - y$

(c) $\dfrac{dx}{dt} = x + 2y - 3$

    $\dfrac{dy}{dt} = -3x + y + 2$

(d) $\dfrac{dx}{dt} = 2x + 3y$

    $\dfrac{dy}{dt} = 3x + 2y$

(e) $\dfrac{dx}{dt} = x + 5y$

    $\dfrac{dy}{dt} = -x - 3y$

(f) $\dfrac{dx}{dt} = 2x - y - 1$

    $\dfrac{dy}{dt} = 2y - 6$

2. A population of cells consisting of a mixture of 2-chromosome and 4-chromosome cells is described approximately by the equations

$$\frac{dT}{dt} = (a - b)T, \qquad \frac{dF}{dt} = bT + cF$$

where T is the number of 2-chromosome cells and F is the number of 4-chromosome cells. The variables T and F clearly cannot be negative, and a, b and c are constants with $a \neq b$ and $c \neq 0$.

Show that whatever the values of a, b and c, the proportion of 2-chromosome cells in

Exercise 9A continued

the population tends to a constant value in the long term, independent of the initial conditions.

Find conditions on the values of a, b and c which ensure that this limiting value of the proportion is non-zero, and find an expression for the limit when this is the case.

3. In a chemical decomposition a compound X produces a compound Y which in turn gives a component Z. These decompositions are governed by the system of differential equations

$$\frac{dx}{dt} = -4x,$$

$$\frac{dy}{dt} = 4x - 2y,$$

$$\frac{dz}{dt} = 2y,$$

where x, y and z are the masses in grams of X, Y and Z respectively, and time t is measured in hours.

Initially $x = 8$, $y = 0$ and $z = 0$.
(i) Find x, y and z in terms of t.
(ii) Determine the maximum value of y.
(iii) Determine the final value of z.

4. In a radioactive decomposition 1 gramme of an unstable isotope X decays into an unstable isotope Y which in turn decays into a stable isotope Z. Assuming the loss of mass in each decomposition is negligible the decomposition process may be modelled by the three simultaneous differential equations

$$\frac{dx}{dt} = -0.5x,$$

$$\frac{dy}{dt} = 0.5x - 0.3y,$$

$$\frac{dz}{dt} = 0.3y$$

where x, y and z are the masses of X, Y and Z respectively at time t.
(i) Solve the first differential equation to find x as a function of time.

(ii) Use the result of part (i) in the second differential equation to find as a function of time.
(iii) Use the results of parts (i) and (ii) to deduce z as a function of time.
(iv) Briefly describe how the masses of X, Y and Z vary with time.

[ME]

5. In a predator–prey environment the rate of growth of the predator population is found to be proportional to the size of the prey population. The rate of change of the prey population, however, is found to depend upon the sizes of both the predator and prey populations. The population dynamics are modelled by assuming that both populations vary continuously. The differential equations governing the relationships between the two populations are

$$100\frac{dx}{dt} = y \qquad \text{and} \qquad 100\frac{dy}{dt} = 2y - x$$

where x and y are the numbers of predator and prey respectively and t is the time in years.

Initially the predator population is 10 thousand and the prey population 5 million.

(i) By eliminating y between the two equations show that the predator population, x, satisfies the second order differential equation

$$10\,000\frac{d^2x}{dt^2} - 200\frac{dx}{dt} + x = 0.$$

(ii) Solve this equation to find the predator population as a function of time.
(iii) Find the prey population, y, as a function of time.
(iv) Determine the size of each population after 5 years.

[ME]

6. For a series of experiments a tank is split into two compartments by a semi-permeable membrane. The first compartment has a volume of V_1 litres and is filled with a solution of a chemical with a concentration of C_1 moles per litre. The second compartment has a volume of V_2 litres and is filled with a solution of the same chemical but with a concentration of C_2 moles per litre. The chemical diffuses through the membrane at a rate proportional to the difference in concentration between the two solutions and in a direction such as to equalise the two concentrations. The differential equations modelling the diffusion process are

$$\frac{dC_1}{dt} = -\frac{k}{V_1}(C_1 - C_2) \text{ and}$$

$$\frac{dC_2}{dt} = \frac{k}{V_2}(C_1 - C_2),$$

where t is the time in hours and k is a constant of diffusion measured in litres per hour.

In a particular experiment $V_1 = 20$, $V_2 = 5$, $k = 0.2$ and, initially, $C_1 = 3$ and $C_2 = 0$.

(i) By eliminating C_2 between the two equations show that C_1 satisfies the second order differential equation

$$\frac{d^2 C_1}{dt^2} + 0.05 \frac{dC_1}{dt} = 0.$$

(ii) Solve this equation to find C_1 as a function of time.

(iii) Find C_2 as a function of time.

(iv) Determine both concentrations after 4 hours, and their final values.

[MEI]

7. The numbers of combatants in each of two armies are denoted, in thousands, by x and y. Each army is reinforced at a rate proportional to its number of combatants and suffers casualties at a rate proportional to the number of combatants in the opposing army. In addition the conflict takes place in a hostile environment which reduces the number of combatants in each army at a constant rate. The pair of simultaneous differential equations modelling the numbers in the conflicting armies are

$$\dot{x} = x - y - 3 \quad \text{and} \quad \dot{y} = y - 4x - 3,$$

where initially $x = 4$, $y = 6$ and t is the time in years.

(i) Write down the initial values of $\dot{x}$ and $\dot{y}$.

(ii) By differentiating the first equation with respect to time and then eliminating $\dot{y}$ from it by using the second equation, show that x satisfies

$$\ddot{x} - 2\dot{x} - 3x = 6.$$

(iii) Find x as a function of t.

(iv) Hence find y as a function of t.

[MEI]

8. The following pair of differential equations provides a simple model for the absorption of a drug by the blood and body tissue. Units are µg and days.

$$\frac{dx}{dt} = -0.05x + 0.01y + 25$$

$$\frac{dy}{dt} = 0.01x - 0.03y$$

where x and y are the amounts of the drug in the blood and the body tissue at time t.

(i) By eliminating y between the two equations, show that x satisfies the second order differential equation

$$\frac{d^2 x}{dt^2} + 0.08 \frac{dx}{dt} + 0.0014x = 0.75.$$

(ii) Solve the equation to find the general solution for x in terms of t.

(iii) Find the general solution for y in terms of t.

(continued overleaf)

Exercise 9A continued

Initially $x = 0$ and $y = 0$.

(iv) Find the particular solutions for x and y in terms of t, and sketch graphs of x and y against t.

(v) Find the equilibrium levels of the drug in the blood and body tissue.

9. Each of two competing species of insect reproduces at a rate proportional to its own number and is adversely affected by the other species at a rate proportional to the number of that other species. At time t, measured in centuries, the populations of these two species are x million and y million. The situation is modelled by the pair of simultaneous equations

$$\frac{dx}{dt} = 2x - 3y \quad \text{and} \quad \frac{dy}{dt} = y - 2x$$

where, initially, at time $t = 0$, $x = 15$ and $y = 10$.

(i) Find the initial rates of change of both x and y.

(ii) Differentiate the first differential equation with respect to t and use this together with the two original differential equations to eliminate y and obtain a second order differential equation for x.

(iii) Solve this second order equation to find x as a function of t.

(iv) Hence find y as a function of t.

(v) Show that one of the species becomes extinct and determine the time at which this occurs.

[MEI]

10. A solution contains substances A, B and C. A chemical reaction changes substance B into substance C at a rate kp where p is the mass of substance C present at time t and k is a constant; simultaneously, a second reaction changes substance A into substance B at a rate of $2kq$, where q is the *combined* mass of substances B and C present. Write down equations for $\dfrac{dp}{dt}$ and $\dfrac{dq}{dt}$.

Initially 48 grams of A, 1 gram of B and 1 gram of C are present. By solving your

equations, show that substance A is exhausted at time t_1, where $kt_1 = \ln 5$, and that the two reactions are completed at time t_2, where $kt_2 = \ln 50$. Find the maximum amount of substance present during the reaction. [SMP

11. In the initial stages of growth a prey population of size x and a predator population of size y are related by the pai of simultaneous differential equations

$$100\dot{x} = 5x - y - 100 \cos t,$$
$$100\dot{y} = 3y + x.$$

By differentiating the first equation with respect to time, eliminate y and its derivatives to form a second-order linear differential equation for x as a function of time.

Find the general solution of this equation Coefficients should be calculated correct two decimal places. [ME

12. A radio-active substance P decays and changes (without loss of mass) into a substance Q, which itself similarly chang into a third substance R. R suffers no further change. The masses of P, Q and R present at time t are given by p, q and r grams respectively. The rates of change are such that

$$\frac{dp}{dt} = -2p \qquad \text{①}$$

and $\qquad \dfrac{dr}{dt} = q.$

Show that $\qquad \dfrac{dq}{dt} = 2p - q.$

Initially (at time $t = 0$) there is one gram o substance P and none of substance Q. Integrate equation ① and hence show tha q satisfies the differential equation

$$\frac{dq}{dt} + q = 2e^{-2t}. \qquad \text{②}$$

Show that ② may be written in the form

$$\frac{d}{dt}(qe^t) = 2e^{-t},$$

and integrate to find q as a function of t. Hence prove that, at any subsequent time there is never more than $\dfrac{1}{2}$ gram of Q present.

[SM

13. (i) Find the equilibrium points of the following systems:

(a)
$$\frac{dx}{dt} = 2x - 3$$
$$\frac{dy}{dt} = x + y + 1$$

(b)
$$\frac{dx}{dt} = x - y + 2$$
$$\frac{dy}{dt} = 3x + y + 2$$

(c)
$$\frac{dx}{dt} = x(x - y)$$
$$\frac{dy}{dt} = y^2 + 3x + 2$$

(Notice that (c) is an example of a non-linear system.)

(ii) Suppose that (x_f, y_f) is the equilibrium point of the linear system

$$\frac{dx}{dt} = a_1 x + b_1 y + c_1$$
$$\frac{dy}{dt} = a_2 x + b_2 y + c_2.$$

Show that the change of variables $X = x - x_f$, $Y = y - y_f$ simplifies the system to

$$\frac{dX}{dt} = a_1 X + b_1 Y$$
$$\frac{dY}{dt} = a_2 X + b_2 Y$$

whose equilibrium point is the origin $(0, 0)$.

(b) Hence show that any autonomous linear system involving two dependent variables can be reduced to a homogeneous linear second order differential equation with constant coefficients in one of the variables.

14. The map below shows three interconnected lakes. The data for each lake is shown in the table

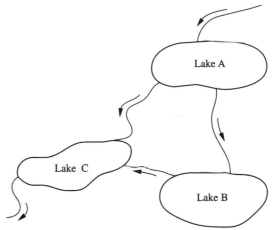

Characteristic	Lake A	Lake B	Lake C
Volume of water (km³)	4871	458	1636
Average outflow rate (km³/year)	209	175	209
Input of chemical pollution (kg/year)	50 000		

A river bearing polluted water flows into Lake A. Water flows directly from Lake A into Lakes B and C and directly from Lake B into Lake C. Water flows from Lake C into the river. Assume that perfect mixing occurs and that the volume of water in each lake remains constant.

(i) Set up a mathematical model for the flow of pollution through the three lakes and find the equilibrium pollution levels in kg l⁻¹, giving your answers to 2 significant figures.

(ii) Starting from an initial pollution–free state, find how long it takes Lake A to reach half of its final pollution level.

(iii) At a time when all the lakes have reached their equilibrium pollution levels, the source of pollution is removed. How long does it take for the pollution in lake A to reach half its original level?

15. Two large tanks on a chemical plant, connected by a series of pipes as in the diagram, each hold 100 litres of water.

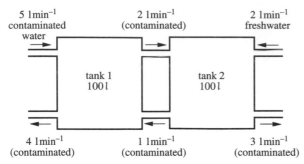

5 lmin⁻¹ contaminated water 2 lmin⁻¹ (contaminated) 2 lmin⁻¹ freshwater

tank 1 100 l   tank 2 100 l

4 lmin⁻¹ (contaminated)   1 lmin⁻¹ (contaminated)   3 lmin⁻¹ (contaminated)

Contaminated water containing 2 kgl^{-1} of a chemical enters tank 1 at a rate of 5 lmin^{-1}. The flow rates of water along the pipes into and out of each tank are shown in the figure. Fresh water also flows into tank 2 at a rate of 2 lmin^{-1}. The liquid in each tank is continuously stirred to ensure uniform concentration of chemical in each tank. Initially the whole system contained fresh water only.

(i) Show that the following pair of differential equations models the amount of chemical in each tank at any time.

$$\frac{dx}{dt} = 10 - 0.06x + 0.01y$$
$$\frac{dy}{dt} = 0.02x - 0.04y$$

where x and y kg are the quantities of the chemical in tanks 1 and 2 respectively at time t minutes.

(ii) Find the equilibrium levels of the chemical in each tank.

(iii) The graph shows the direction field for the differential equation

$$\frac{dy}{dx} = \frac{0.02x - 0.04y}{10 - 0.06x + 0.01y}$$

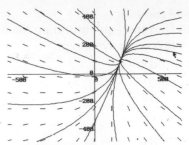

Describe what happens to the amount of the chemical in each tank if at some instant the values of x and y correspond to the co-ordinates of point A. Repeat this for points B, C and D.

16. The following pair of differential equations is proposed as an economic model for the price P of a single item in a market where Q is the number of the item available:

$$\frac{dP}{dt} = aP\left(\frac{b}{Q} - P\right)$$
$$\frac{dQ}{dt} = cQ(fP - Q)$$

where a, b, c and f are positive constants.

(i) Consider the case $a = 1, b = 20\,000$, $c = 1$ and $f = 30$.
  (a) Find the equilibrium points for this economic model.
  (b) By drawing the direction field and some solution curves, investigate the model for different initial conditions.

(ii) Suggest the assumptions and simplifications upon which this model might be based.

17. The following pair of differential equations is a simple predator-prey model for the population of foxes and rabbits in a park in Reading.

$$\frac{dR}{dt} = R(120 - 01.R - 2F)$$
$$\frac{dF}{dt} = F(80 - 0.1R + 2F).$$

Find the equilibrium populations of foxes and rabbits, and describe what happens to the number of foxes and rabbits for different initial populations.

Investigation

A cascade of water tanks

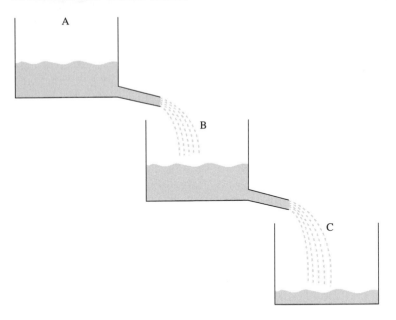

The diagram shows three tanks. Initially tank A is full of water and tanks B and C are empty. In the final state tanks A and B are empty and tank C is full.

The amount of water in tank B varies between the start and the finish. Construct a mathematical model to predict when tank B will contain its maximum volume, and what that maximum volume will be.

(**Note:** Before you construct your model you will need to find out, either by experiments or other research, the relationship between the flow rate out of a tank and the depth of water in it.)

KEY POINTS

- A system of differential equations involves two or more dependent variables and one independent variable.
- In the linear system

$$\frac{dx}{dt} = a_1 x + b_1 y + f_1(t) \qquad ①$$

$$\frac{dy}{dt} = a_2 x + b_2 y + f_2(t) \qquad ②$$

x and y are the dependent variables, t the independent variable.

KEY POINTS *continued*

- Solving such a system of equations involves finding x in terms of t and y in terms of t.
- The relationship between x and t may be found by:
  (i)   differentiating equation ①;
  (ii) substituting for $\dfrac{dy}{dt}$ from equation ② and for y from equation ①;
  (iii) solving the resulting second order linear equation.

  The relationship between y and t may then be found by substituting for $\dfrac{dx}{dt}$ and x in equation ①.

- In the case when $f_1(t) = 0$ and $f_2(t) = 0$ the resulting second order equation is homogeneous; otherwise it is non-homogeneous.
- The graph of y against x is called the solution curve.
- If the solution reaches an equilibrium point it does not move away from it. It may alternatively approach an equilibrium point as $t \to \infty$.
- Non-linear systems may be investigated by:
  (i)   drawing the tangent field;
  (ii) investigating equilibrium points;
  (iii) using a step-by-step method of solution.

Answers

Exercise 1A

1. $\dfrac{dm}{dt} = -km$

2. $\dfrac{dv}{dt} = -\dfrac{v^2}{400}$

3. $\dfrac{dP}{dt} = \dfrac{P}{34}$

4. $\dfrac{dP}{dt} = \dfrac{P}{6}$

5. $\dfrac{dV}{dt} = -2\sqrt{h}$

6. $\dfrac{dN}{dt} = 2\sqrt{N}$

7. $\dfrac{dC}{dt} = -0.2C$

8. $\dfrac{d\theta}{dt} = \dfrac{20-\theta}{120}$

9. (i) $\dfrac{dA}{dt} = kr$ (ii) $\dfrac{dA}{dr} = 2\pi r$

(iii) $\dfrac{dr}{dr} = \dfrac{k}{2\pi}$

10. $\dfrac{dr}{dr} = -\dfrac{1}{100\pi}$

11. $\dfrac{d\theta}{dx} = -\dfrac{16}{25}x$

12. $\dfrac{dp}{dh} = \begin{cases} 9800(1+0.001h) & 0 \le h \le 100 \\ 10780 & h > 100 \end{cases}$

13. $\dfrac{dy}{dx} = \dfrac{b-y}{a+vt-x}$

14. $\dfrac{dh}{dt} = \dfrac{4.8}{\pi h^2}$

15. $\dfrac{dV}{dt} = -\sqrt{20h}$, $\dfrac{dh}{dt} = -\dfrac{\sqrt{5h}}{2}$

16. $\dfrac{d\theta}{dt} = k(22-\theta)$

Exercise 1B

3. (ii) 0
(iii) Tends to 20
4. (iii) Coffee $T = 20 + 60e^{-0.1t}$
Juice $T = 20 - 18e^{-0.1t}$

5. (ii) $P = 50e^{t/20}$
(iii) $P = 100e^{t/20}$
(iv)

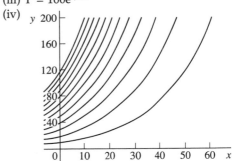

6. (ii) 69.3
(iii) 20
(iv) $m = 50e^{-t/100}$

8. $y = x^3 - \dfrac{x^2}{2} + x + 2.5$

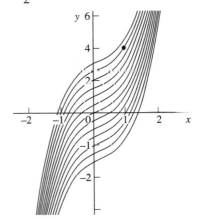

9. $s = 4t - 5t^2 + c$
$s = 4t - 5t^2 + 11$
10. A and E
11. (iii) $A = P_0 e^{-kt_0}$
12. (ii) $\alpha + A, \alpha$
(iii) $\alpha = 25, A = 65$
13. (ii) 4

Exercise 2A
1. (i)

y＼x	-2	-1	0	1	2
2	-4	-2	0	2	4
1	-4	-2	0	2	4
0	-4	-2	0	2	4
-1	-4	-2	0	2	4
-2	-4	-2	0	2	4

(ii), (iii)

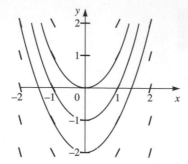

(iv) $y = x^2 + c$

(v) $y = x^2 - 2$, $y = x^2 - 1$, $y = x^2$.

(vi) Isoclines are straight lines parallel to the y axis.

2. (i)

y \ x	-2	-1	0	1	2
2	1.5	1	0.5	0	-0.5
1	3	2	1	0	-1
0	∞	∞	∞	$-$	∞
-1	-3	-2	-1	0	1
-2	-1.5	-1	-0.5	0	0.5

(ii)/(iii)

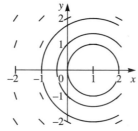

(iv) $y = 2 - 2x$

(v) Isoclines are straight lines through the point $(1, 0)$ which is the common centre of the family of circles.

3. Isoclines are straight lines parallel to x axis

(ii)/(iii)

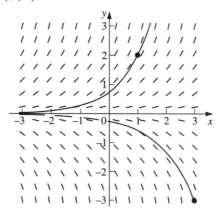

4. (i) Isoclines are straight lines parallel to x axis, $y = 1$.

(ii)/(iii)

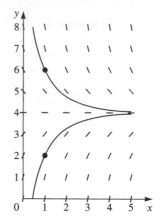

(iv) Each is the reflection of the other in the line $y = 4$.

5. (i)/(ii)

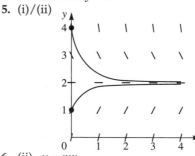

6. (ii) $y = mx$

(iii)
(c) the y axis

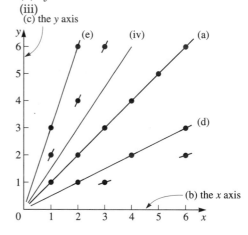

(iv) All straight lines coming from the origin.

(v) Gradient $= \frac{0}{0}$ which is undefined.

7. (i) $y = m - x$

(ii)/(iii)/(iv)

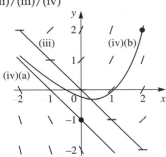

8. (i) $y = m + x$

(ii)/(iii)/(iv)

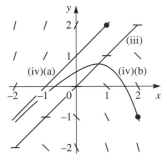

9. (i) $y = m/x$ rectangular hyperbola

(ii), (iv)

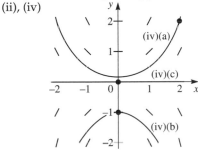

(iii) $xy = 0$ i.e. the x and y axes

10. (i)//(ii) see graph (iii) $i = 3$

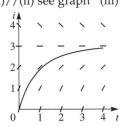

11. The population is 12 million during 1993.

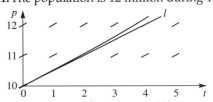

12. (i), (ii) $v(\text{ms}^{-1})$

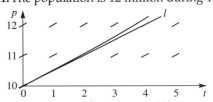

(iii) Terminal speed is 8 ms^{-1}.

Exercise 3A

1. (i) $y = Ae^x$ **(ii)** $y = Ae^{\frac{1}{2}x^2}$

(iii) $y = Ae^{\frac{1}{4}x^4}$ **(iv)** $y = \ln(x^3 + c)$

(v) $\ln|1 + y| = \dfrac{x^2}{2} + c$ or $y = Ae^{x^2/2} - 1$

(vi) not possible **(vii)** $e^x + e^{-y} = c$

(viii) $y = \dfrac{A(x-1)}{x}$ **(ix)** not possible

(x) $y^3 = K - 3\cos x$ **(xi)** $y = A(x-2) - 2$

(xii) $\ln\left|\dfrac{x-8}{x}\right| = 8t + c$ or $x = \dfrac{8}{1 - Ae^{8t}}$

2. (i) $y = \dfrac{10}{1 - 10\ln x}$ **(ii)** $y = \dfrac{2}{1 - 2\ln x}$

(iii) $y^2 = \dfrac{2x^3 + 298}{3}$ **(iv)** $y = \ln\left(\dfrac{3 - \cos 2x}{2}\right)$

(v) $y = \ln\left(e^{10} + \dfrac{x^3}{3}\right)$ **(vi)** $e^{-y}(1 + y) = 3c^{-2} - x$

3. (i) $m = Ac^{-5t}$ **(ii)** $m = 10e^{-5t}$

4. (i) $P = Ae^{0.7t}$ **(ii)** $P = 100e^{0.7t}$

(iii) 0.99 minutes

5. (i) $v = \dfrac{20}{1 + 2t}$ (ii) 4.5 seconds

6. (i) $h = \dfrac{1}{\dfrac{\pi t}{4} + c}$ or $h = \dfrac{4}{\pi t + k}$

(ii) $t = \dfrac{4}{9\pi H}$

7. (i) $h = \left(2 - \dfrac{\sqrt{5}}{4}t\right)^2$

(ii) $t = \dfrac{8}{\sqrt{5}}$ (≈ 3.58) minutes to empty.

8. (i) $|10-2v| = Ae^{-0.2t}$

(ii) a) $v = 50(1-e^{-0.2t})$, 50

b) $v = 50 + 30e^{-0.2t}$, 50

9. (i) $T = 20 + 80e^{-0.5t}$

(ii) $t = 1.96$ minutes

10. $v = \dfrac{40e^{-0.2t}}{41 - 40e^{-0.2t}}$

11. (i) $i = 0.2 - Ae^{-2500t}$

(ii) $i = 0.2(1 - e^{-2500t})$

(iii) $i = \dfrac{V}{R} - Ae^{-Rt/L}$

12. (ii) $t = 62.5\ln\left(\dfrac{2.5}{2.5-h}\right)$

(iii) It takes an infinite time to fill the lock.

(iv) 446 seconds (to nearest second).

13. (i) $v(0) = 0$

(iii)

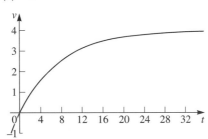

(iv) $t = 11.09$ seconds

14. 6 minutes before midnight.

15. (ii) $\dfrac{dC}{dt} = -0.004C$

(iii) 101 minutes

16. (i) $\dfrac{mg}{V}$

(iv) Air resistance reduces the maximum height by approximately $\dfrac{U^3}{3gV}$

17. (i) $\dfrac{dh}{dt} = kh$, k is constant   (ii) $h = 8e^{0.56t}$

(iii) The exponential model suggests that sweet peas grow forever. The data show a limiting height.

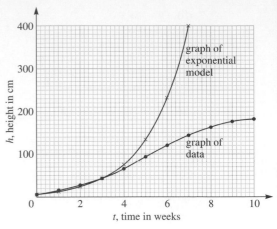

graph of exponential model

graph of data

(iv) When $h = H$ we have $\dfrac{dh}{dt} = 0$, i.e. an equilibrium height.

(v) $H = 190.8$, $A = 0.0438$, $a = 0.616$

$$h = \dfrac{190.8}{1 + 22.83e^{-0.616t}}$$

(vi)

t	0	1	2	3	4	5
h	8.0	14.3	24.9	41.5	64.8	93.1

t	6	7	8	9	10
h	121.8	146.1	163.7	175.2	182.0

The logistic equation models the plant growth extremely well.

18. (i) $\dfrac{dS}{dt} = -kS(10^5 - S)$

(iii) $I = \dfrac{10^5}{1 + 99e^{-0.321t}}$, $S = \dfrac{9.9 \times 10^6 e^{-0.321t}}{1 + 99e^{-0.321t}}$

(v) 14.3 weeks.

Exercise 4A

1. (i) $e^{x^3/3}$ (ii) $e^{-\cos x}$

(iii) $x^{-\frac{1}{4}}$ (iv) x

(v) e^{7x} (vi) $\sec x$

2. (i) $y = \dfrac{x^2}{4} - \dfrac{1}{4x^2}$ (ii) $y = 4 - 2e^{-x^2/2}$

(iii) $y = 3e^{-3(x^2-1)}$ (iv) $y = \dfrac{1}{2}(3e^{x^2} - 1)$

(v) $y = x^3 - x^2$ (vi) $y = \dfrac{1}{3}(1 - 4e^{-x^3})$

3. (i) $v = 25 + Ae^{-0.4t}$ (ii) $v = 25(1 - e^{-0.4t})$

(v) The method of separation of variables is usually preferred as it involves less work.

4. (i) $k = \frac{1}{3}$

(ii) $\frac{dv}{dt} = 10 - \frac{1}{3}v$

(iii) $v = 30 + Ae^{-\frac{1}{3}t}$

(iv) $v = 30(1 + e^{-\frac{1}{3}t})$

5. (i) $y = Ae^{-k_2 t} + \frac{k_1 a}{(k_2 - k_1)}e^{-k_1 t}$

(ii) $y = \frac{k_1 a}{(k_2 - k_1)}(e^{-k_1 t} - e^{-k_2 t})$

(iii) $\frac{dx}{dt} = -k_1 x, \quad \frac{dz}{dt} = k_2 y$

(iv) $y = kate^{-kt}$ where $k_1 = k_2 = k$.

Exercise 5A

1. 1.6319 with $h = 0.2$
1.6252 with $h = 0.1$

2. 2.7387 with $h = 0.1$
2.7466 with $h = 0.05$

3. 2.2666 with $h = 0.2$
2.2745 with $h = 0.1$

4. 1.7680 with $h = 0.1$
1.8871 with $h = 0.05$

Exercise 5B

1. (i)

h	$y(6)$
1.0	96
0.5	96.8
0.2	97.2672
0.1	97.4209

(ii) 97.5829

(iii) $y = 1280e^{-0.05(x-5)} + 80x - 1600$
$y(6) = 97.5737$ (to 4 d.p.)

(iv) $h \le 0.00032$
3125 steps

2. (i)

h	$y(2)$
0.2	1.1952
0.1	1.4205
0.05	1.8131
0.01	1.8982

(ii) 1.9324

(iii) $y(2) = 2$

(iv) $h \le 4.9 \times 10^{-6}$
200 000 steps

3. 3.0984

4. 2.8430

5. (i) 0.22
(ii) 0.24

6. (i) (a) $k_1 = 0.025$
(b) $k_2 = 0.000125$
(c) $k_3 = 0.00559$

(ii) $v = 20(1 - e^{-0.5t})$

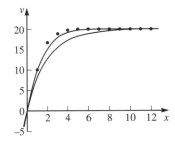

7.

Exercise 6A

1. (i) $y = Ae^{3x}$

(ii) $y = Ae^{-7x}$

(iii) $x = Ae^{-t}$

(iv) $p = Ae^{0.02t}$

(v) $z = Ac^{0.2t}$

2. (i) $y = 3e^{-2x}$

(ii) $y = e^{5x/2}$

(iii) $x = 2e^{\frac{1}{3}(1-t)}$

(iv) $p = p_0 e^{kt}$

(v) $m = m_0 e^{kt}$

3. $y = Ae^{-5x}$

4. (i) all three, $y = Ae^{17x}$

(ii) separation of variables only,
$$y = -\frac{1}{x + c}$$

(iii) separation of variables or integrating

factor, $y = Ae^{-\frac{x^2}{2}}$

(iv) all three, $y = Ae^{3x}$

(v) all three (after division by y),
$y = Ae^x$ (which includes $y = 0$)

5. (i) $h(x, y)$ is a product (or quotient) of a function of x alone and a function of y alone

(ii) $h(x, y)$ is linear in y

(iii) $h(x, y) = cy$ for some constant c

Exercise 6B

1. $y = Ae^{2x} + Be^x$

2. $y = Ae^{-x/3} + Be^{-x/2}$

3. $y = Ae^{x/2} + Be^{-x}$

4. $y = Ae^{-5x} + Be^x$

5. $y = Ae^{2x} + Be^{-2x}$

6. $y = Ae^x + Be^{2x} + Ce^{3x}$

7. $x = Ae^{-3t} + Be^{2t}$

8. $x = Ae^{3t} + Be^{-3t}$

9. $u = A + Be^{3t}$

10. $x = Ae^{\omega t} + Be^{-\omega t}$ where $\omega = \sqrt{\dfrac{k}{m}}$

Exercise 6C

1. $y = -2e^{3x} + 3e^{2x}$

$y \to -\infty$ as $x \to \infty$

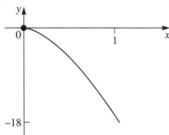

2. $y = \dfrac{1}{6}(e^{3x} - e^{-3x})$ $y \to \infty$ as $x \to \infty$

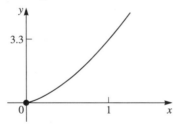

3. $x = \dfrac{2e}{1 - e^{-4}}(e^{-t} - e^{-5t})$ $x \to 0$ as $t \to \infty$

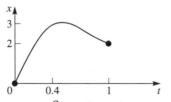

4. $v = \dfrac{e}{e^2 - 1}(e^t - e^{-t})$ $v \to \infty$ as $t \to \infty$

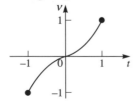

5. $y = \dfrac{4}{5}(e^{5x} - 1)$ $y \to \infty$ as $x \to \infty$

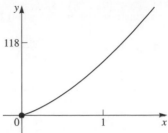

6. (i) $T = Ae^{2x} + Be^{-2x}$

(ii) $T = -52.3e^{2x} + 152.3e^{-2x}$

(iii) 87.9 °C

Exercise 6D

1. $x = \dfrac{1}{3}\sin 3t$ oscillating system, amplitude $\dfrac{1}{3}$, period $\dfrac{2\pi}{3}$

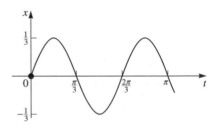

2. $x = 4\cos t/2$, oscillating system, amplitude 4, period 4π

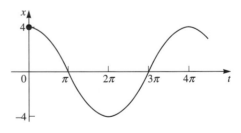

3. $x = \sin 2t$ oscillating system, amplitude 1, period π

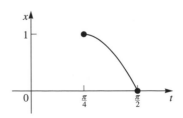

4. $x = -9.47 \sin 2\sqrt{3}t$, oscillating system,

amplitude 9.47, period $\dfrac{\pi}{\sqrt{3}}$

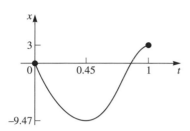

5. $x = \sqrt{\dfrac{m}{k}} \sin \sqrt{\dfrac{k}{m}}\, t$

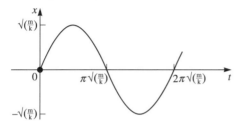

6. (i) $y = e^{2x}(A\sin x + B\cos x)$
   (ii) $y = e^{x}(A\sin 2x + B\cos 2x)$

   (iii) $x = a^{-t}(A\sin\sqrt{3}t + B\cos\sqrt{3}t)$

   (iv) $x = e^{-t/2}(A\sin t + B\cos t)$

7. (i) $y = 2e^{-x}\sin x$ decaying oscillations

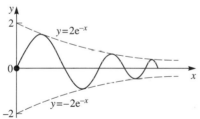

   (ii) $y = e^{x}\left(2\cos\dfrac{1}{2}x - 4\sin\dfrac{1}{2}x\right)$

   $= \sqrt{20}e^{x}\ \cos(\tfrac{x}{2} + 1.11)$

   increasing oscillations

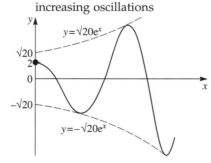

(iii) $y = 3e^{-x + \pi/4}\sin 2x = 6.58e^{-x}\sin 2x$

decaying oscillations

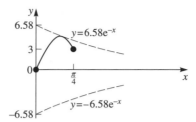

(iv) $\quad x = e^{\frac{3}{2}t}\left(\cos\dfrac{\sqrt{7}t}{2} - \dfrac{3}{\sqrt{7}}\sin\dfrac{\sqrt{7}t}{2}\right)$

$= 1.51e^{\frac{3}{2}t}\cos\left(\dfrac{\sqrt{7}}{2}t + 0.85\right)$

increasing oscillations.

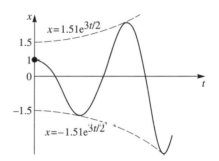

Exercise 6E

1. (i) $y = Ae^{4x} + Be^{-4x}$
   (ii) $x = A\sin\omega t + B\cos\omega t$
   (iii) $y = Ae^{1.79x} + Be^{-2.79x}$

   (iv) $x = e^{-\frac{t}{6}}\left(A\sin\dfrac{\sqrt{23}}{6}t + B\cos\dfrac{\sqrt{23}}{6}t\right)$

   (v) $y = e^{4x}(Ax + B)$
   (vi) $x = Ae^{-1.71t} + Be^{-0.29t}$

   (vii) $y = A + Be^{-\frac{2}{7}x}$

   (viii) $y = e^{\frac{2}{3}x}(Ax + B)$
   (ix) $r^2 > 4\,km$

   $x = Ae^{(\alpha+\beta)t} + Be^{(\alpha-\beta)t}\begin{cases}\alpha = -\dfrac{r}{2m} \\[2mm] \beta = \dfrac{\sqrt{(r^2 - 4km)}}{2m}\end{cases}$

   $r^2 = 4km \qquad x = e^{-rt/(2m)}(At + B)$

$r^2 < 4km$

$$x = e^{-\frac{rt}{2m}}\left(A\sin\frac{\sqrt{(4km - r^2)}}{2m}t + B\cos\frac{\sqrt{(4km - r^2)}}{2m}t \right)$$

(x) $y = Ae^{2x} + Be^{-2x} + C\sin 2x + D\cos 2x$

3. (i) $x = \dfrac{\sqrt{2}}{4}\sin 2\sqrt{2}t + \cos 2\sqrt{2}t$

$\qquad = \dfrac{3\sqrt{2}}{4}\sin(2\sqrt{2}t + 1.23)$

oscillations with constant amplitude

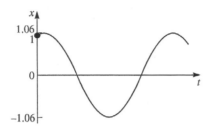

(ii) $x = \dfrac{4}{5}(1 - e^{-5t})$ Moves quickly
initially towards its limiting position
position $x = 0.8$

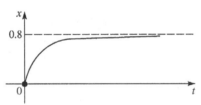

(iii) $x = e^{t/2}\left(\cos\dfrac{\sqrt{3}}{2}t - \dfrac{1}{\sqrt{3}}\sin\dfrac{\sqrt{3}}{2}t \right)$

$\qquad = \dfrac{2}{\sqrt{3}}e^{t/2}\cos\left(\dfrac{\sqrt{3}}{2}t + 0.52 \right)$

Exponentially increasing oscillations

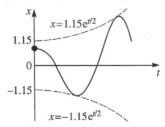

4. (i) $y = 2xe^{1-x}$ (ii) $x = (1 - 3t)e^{3t}$

(iii) $x = 2(2 - t)e^{-\frac{1}{2}}$ (iv) $y = 2xe^{-kx}$

Exercise 6F

1. $y = -2x - 5$

2. $x = \dfrac{1}{4}t + \dfrac{1}{2}$

3. $y = \dfrac{5}{3}x + \dfrac{2}{9}$

4. $x = \dfrac{l}{\omega^2}$

5. $v = -\dfrac{1}{2}t - \dfrac{3}{2}$

6. $y = \dfrac{1}{2}x^2 - 7x$

7. $y = -0.08\cos 3x + 0.06\sin 3x$

8. $y = -\dfrac{1}{289}(8\cos 4x + 15\sin 4x)$

9. $v = -0.2\cos t - 0.6\sin t$

10. $x = 0.6\cos 2t$

11. $x = -\dfrac{1}{4 + \omega^2}\sin \omega t$

12. $x = \dfrac{1}{2}\cos t + \dfrac{1}{4}\sin t$

13. $y = \dfrac{1}{5}e^{-x}$

14. $x = \dfrac{3}{4}e^{-2t}$

15. $y = -5e^{2x}$

16. $x = \dfrac{1}{12}e^{-3t}$

Exercise 6G

1. (i) $y = A\cos 2x + B\sin 2x + \dfrac{1}{5}e^x + \dfrac{1}{8}e^{2x}$

(ii) $x = Ae^{3t} + Be^{-t} + e^{-2t} + 2e^{4t}$

(iii) $y = A\cos\sqrt{2}x + B\sin\sqrt{2}x + e^x + \sin x$
(iv)

$y = e^{-\frac{1}{2}x}\left(A\cos\dfrac{\sqrt{3}}{2}x + B\sin\dfrac{\sqrt{3}}{2}x\right) + 4 + \sin x$

(v)

$y = e^{-\frac{1}{2}x}\left(A\cos\dfrac{\sqrt{3}}{2}x + B\sin\dfrac{\sqrt{3}}{2}x\right) + 3x - 1 + \dfrac{1}{3}$

(vi) $y = A\cos x + B\sin x - \dfrac{1}{2}x\cos x$

(vii) $y = Ae^{-4x} + Be^x + \dfrac{1}{5}xe^x$

(viii) $y = Ae^x + Be^{3x} - xe^x + \dfrac{1}{2}xe^{3x}$

(ix) $y = (Ax + B)e^{3x} + 2x^2e^{3x}$

(x) $y = A + Be^{2x} - \dfrac{1}{4}x^2 - \dfrac{1}{4}x$

2. (i) $y = -6e^{3x} + e^{2x} + 6x + 5$

(ii) $x = -2\cos 3t + \sin 3t + 2e^{-t}$

(iii) $y = e^{-x}(-\cos 2x + 1)$

(iv) $x = -e^{-2t} + 4e^t - \cos 2t - 3\sin 2t$

(v) $y = 2 - e^{-x} + \dfrac{1}{2}x^2 - x$

(vi) $x = \dfrac{1}{2}e^{2t} - e^t + \dfrac{1}{2} + te^t$

(vii) $y = 2\sin 2x - 3x\cos 2x$

(viii) $y = e^{-2x}(2\cos x + 3\sin x) - \cos x + \sin x$

(ix) $y = \dfrac{1}{2}e^{-2x} + \dfrac{1}{2}e^{2x} + x - 1$

(x) $\dfrac{1}{2}(x+1)^2 e^{-2x}$

3. (i) $p = Ae^{-t}$

(ii) $p = 100 - 25(\cos t - \sin t)$

(iii) $p = 100 - 25(\cos t - \sin t) - 55e^{-t}$

(iv) 100

(v) $25\sqrt{2}$

4. (i) $\dfrac{dv}{dx} + v\tan x = e^{-x}\cos x$

(ii) $v = -e^{-x}\cos x + \dfrac{3}{2}\cos x$

$y = \dfrac{1}{2}e^{-x}(\cos x - \sin x) + \dfrac{3}{2}\sin x + \dfrac{1}{2}$

(iv) $\dfrac{3\sqrt{2}}{4} + \dfrac{1}{2}$

Exercise 7A

1. (i) $\dfrac{d^2x}{dt^2} + 64x = 0$ $\dfrac{dx}{dt} = 0, x = 0.1$ when $t = 0$

(ii) $x = 0.1\cos 8t$

(iii) Period $= \dfrac{\pi}{4}$s , amplitude $= 0.1$m

2. (i) 0.357 m

(ii) $\dfrac{d^2x}{dt^2} + 62.5x = 0$

$\dfrac{dx}{dt} = 0, x = 0.1$ when $t = 0$

(iii) $x = 0.1\cos\sqrt{62.5}t$

(iv) Period $= \dfrac{2\pi}{\sqrt{62.5}} = 0.79s$,

amplitude $= 0.1$ m (10 cm)

3. (i) 0.039m

(ii) $l = 0.439 + 0.03\cos\sqrt{250}t$

4. period $= \dfrac{2\pi}{\sqrt{250}} = 0.397s$

5. (i) $\dfrac{5\pi}{180} = 0.0873$ radians

(ii) period $= 3.48s$

(iii) $a = 0.087$

$\omega = 1.807$

$\varepsilon = 0$

(iv) maximum speed $= 0.474$ ms^{-1}

6. (i) length $= \dfrac{g}{\pi^2} = 0.993$ m.

(ii) (a) no change

(b) increases the period by a factor $\sqrt{2}$

(c) no change

(d) increases the period by a factor $\sqrt{6}$

7. (i) $\dfrac{d^2\theta}{dt^2} = -\dfrac{g}{4}\sin\theta$

neglect air resistance

inextensible string

no friction at support

small angle vibrations ($\sin\theta \simeq \theta$)

bob is a particle

(ii) period $= 4.01$ s

8. (i) net force $= 0$ and only forces when in equilibrium are

F and mg. constant $= \dfrac{mg}{l}$

(ii) $F = \dfrac{mg}{l}(l+x), \dfrac{d^2x}{dt^2} + \dfrac{g}{l}x = 0$;

(iii) $2\pi\sqrt{\dfrac{l}{g}}$

9. (i) $V = \dfrac{2\pi}{3} = 2.09$ ms^{-1}

(ii) 0.87 m

(iii) 0.022 N

Exercise 7B

1. (i) $\dfrac{d^2x}{dt^2} + 80x = 0$

$\dfrac{dx}{dt} = 0$ and $x = 0.1$ when $t = 0$.

(ii) $x = 0.1\cos\sqrt{80}t$.

(iii)

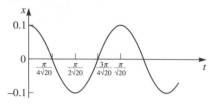

(iv) For critical damping, damping constant $= \sqrt{20}\ (\simeq 4.47)$

(v) for m = 0.3, underdamped system for m = 0.2, overdamped system

2. (i) $x = e^{-\frac{3}{2}t}A\cos(\frac{\sqrt{31}}{2}t + \varepsilon) + 0.05$

(ii) $x = e^{-\frac{3}{2}t}0.057\cos(2.78t - 0.49) + 0.05$

$x \to 0.05$ as $t \to \infty$

(iii) 1.16

3. (i) $q = (2 + 9t)e^{-2.5t}$

(ii)

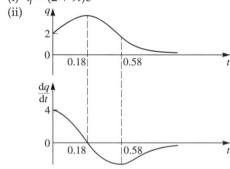

(iii) $q \to 0$, and $= \dfrac{dq}{dt} \to 0$ as $t \to \infty$

4. (i) $T = 37.3 + 12.7e^{-0.5t}$

(ii) $T = 42.0\ ^\circ C$

(iii)

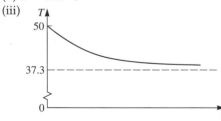

(iv) 37.3 $^\circ C$

5. (i) $\theta = e^{-2t}A\cos(t + \varepsilon)$

(ii) $1.76e^{-2t}\cos(t - 1.11)$

(iii) The motion effectively ends at about t = 2.5. Although θ subsequently does become negative. The amplitude of the oscillation is very small and decreases rapidly.

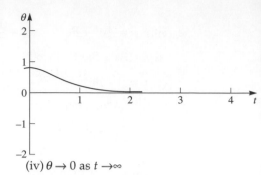

(iv) $\theta \to 0$ as $t \to \infty$

6. (i) $10^4\dfrac{d^2x}{dt^2} + 2\times10^5\dfrac{dx}{dt} + 10^6x = 0$

where x is depression of spring from its natural length

$\dfrac{dx}{dt} = 5$, $x = 0$ when $t = 0$, $x = 5te^{-10t}$

(ii)

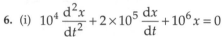

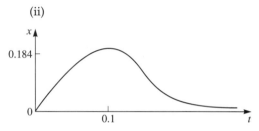

The buffer is compressed 18.4 cm after 0.1s and is then gradually released.

7. (i) $\underline{m = 2}$ $x = 4 - \dfrac{20}{3}e^{-t} + \dfrac{8}{3}e^{-2.5t}$

$\underline{m = 2.45}$ $x = 4.9 - (7.01t + 4.9)e^{-1.43t}$

$\underline{m = 4}$ $x = 8 - e^{-0.875t}(10.0\sin0.70t + 8\cos0.70t)$

$\qquad = 8 - 12.8e^{-0.875t}\cos(0.70t - 0.90)$

(ii)

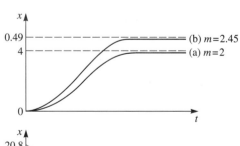

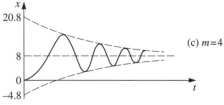

8. (i) $p = \dfrac{4m}{3\pi a^2 b}$

(ii) $m\dfrac{d^2 x}{dt^2}$ = mass multiplied by

downward acceleration

$2mk\dfrac{dx}{dt}$ = resistance to motion

$\dfrac{4mg}{3b}x$ = extra upward thrust of $\pi a^2 x \rho g$

from displaced water whose volume is $\pi a^2 x$.

(iv) $x = e^{-kt}\left(\dfrac{b}{4}\cos\omega t + \dfrac{kb}{4\omega}\sin\omega t\right)$

9. (ii) $x = A\sin\omega t + B\cos\omega t + \dfrac{c}{\omega^2}$.

$\dfrac{c}{\omega^2}$ is the value of x when the system is in equilibrium.

(iii) (a) $r > 2\omega$, $x = Ae^{\lambda_1 t} + Be^{\lambda_2 t} + \dfrac{c}{\omega^2}$

where λ_1, λ_2 are the real roots of $\lambda^2 + r\lambda + \omega^2 = 0$.

(b) $r = 2\omega$, $x = (At + B)e^{-rt/2} + \dfrac{c}{\omega^2}$

(c) $r < 2\omega$,

$x = e^{-\frac{rt}{2}}(A\sin\Omega t + B\cos\Omega t) + \dfrac{c}{\omega^2}$

where $\Omega = \dfrac{\sqrt{(4\omega^2 - r^2)}}{2}$.

Exercise 8A

1. (ii) $x = 0.5 + \dfrac{1}{50 - \Omega^2}(5\cos\Omega t - 0.1\Omega^2\cos\sqrt{50}t)$

(iii)

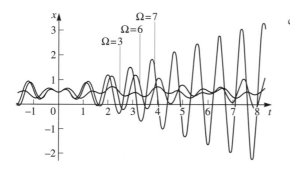

(iv) $\Omega = \sqrt{50}$ for resonance

$x = 0.5 + \dfrac{t}{2\sqrt{2}}\sin\sqrt{50}t + 0.1\cos\sqrt{50}t$

2. (i) $q = \dfrac{10(100\sin\Omega t - \Omega\sin 100t)}{10\,000 - \Omega^2}$

(ii) $\Omega = 100$ for resonance

$q = 0.05\sin 100t - 5t\cos 100t$

$t \approx 2000s$ (636.63π)

(iii)

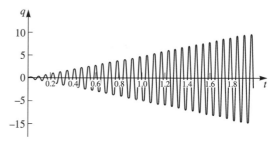

3. (i) $y = A\sin\omega t + B\cos\omega t + \dfrac{F\omega^2}{\omega^2 - p^2}\sin pt$

$(p \neq \omega)$

(ii) $y = \cos\omega t + \dfrac{\omega F}{(\omega^2 - p^2)}(\omega\sin pt - p\sin\omega t)$

(iii)

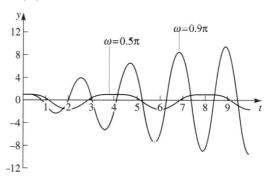

Exercise 8B

1. (ii) $x = e^{-1.5t}(-0.05\cos 2.78t + 0.009\sin 2.78t)$
$+ 1 + 0.05\cos 2t - 0.05\sin 2t$

(iii)

DE

2. (i) $I_c = Ae^{-2t}$

(ii) $I_p = -\dfrac{3}{13}\cos 3t + \dfrac{2}{13}\sin 3t$

(iii) $I = \dfrac{3}{13}e^{-2t} - \dfrac{3}{13}\cos 3t + \dfrac{2}{13}\sin 3t$

(iv) $\dfrac{1}{\sqrt{13}} \approx 0.277$

3. (i) $e^{-t}(A\sin 3t + B\cos 3t)$
(ii) underdamped system

(iii) $\dfrac{(10-\omega^2)\sin\omega t - 2\omega\cos\omega t}{(10-\omega^2)^2 + 4\omega^2}$

4. (i) $e^{-50t}(A\sin 50\sqrt{3}t + B\cos 50\sqrt{3}t)$

(ii) underdamped
(iii) $-0.1\cos 100t$,

$q = e^{-50t}(A\sin 50\sqrt{3}t + B\cos 50\sqrt{3}t) - 0.1\cos 100t$

(iv) $A = \dfrac{1}{10\sqrt{3}}$, $B = \dfrac{1}{10}$

(v)

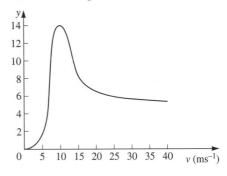

5. (i) $x = B\sin\omega t + C\cos\omega t$
(ii) damping term, forcing term

(iii) $x = B\sin\omega t + C\cos\omega t + \dfrac{A}{\omega^2 - \alpha^2}\cos\alpha t$

(iv) The closer α gets to ω, the larger the amplitude of vibrations becomes

6. (i) $x = \dfrac{m^2 g}{a^2 r^2}\left(e^{-\frac{rat}{m}} - 1\right) + \dfrac{mg}{ar}t$

$v = \dfrac{mg}{ar}\left(1 - e^{-\frac{rat}{m}}\right)$

(ii)

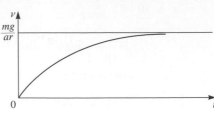

Speed of sphere increases until it

reaches its terminal velocity $\dfrac{mg}{ar}$.

7. (iii) Amplitude =

$$\dfrac{0.05\Omega^2}{\sqrt{\left[\left(100-\Omega^2\right)^2 + (11.25\Omega)^2\right]}}$$

where $\Omega = \pi v$

(iv) Avoid a speed of 10 ms^{-1}

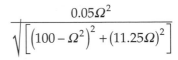

(v) Increase amplitude of steady state solution

8. (i) $k = 1$
(ii) $x = 1 - (t+1)e^{-t}$
(iii) $x = 1 - e^{-0.6t}(0.75\sin 0.8t + \cos 0.8t)$
(iv) $1 + e^{-0.75\pi} = 1.095$.

Exercise 9A

1. (a) (i) $x = Ae^{2t} + Be^t$
$y = Ae^{2t} + 2Be^t$
(ii) $x = e^t$, $y = 2e^t$
(iii) $(x, y) \to \infty$ as $t \to \infty$ along the line $y = 2x$

(b) (i) $\quad x = Ae^{\sqrt{2}t} + Be^{-\sqrt{2}t}$

$$y = (\sqrt{2} - 1)Ae^{\sqrt{2}t} - (\sqrt{2} + 1)Be^{-\sqrt{2}t}$$

(ii) $\quad x = 1.56e^{\sqrt{2}t} - 0.56e^{-\sqrt{2}t},$

$$y = 0.65e^{\sqrt{2}t} + 1.35e^{-\sqrt{2}t}$$

(iii) $(x, y) \to \infty$ as $t \to \infty$ along the line $y = 0.42x$

(c) (i) $\quad x = e^t(A\sin\sqrt{6}t + B\cos\sqrt{6}t) + 1$

$$y = \frac{\sqrt{6}e^t}{2}(-B\sin\sqrt{6}t + A\cos\sqrt{6}t) + 1$$

(ii)

$$x = 1 + \frac{\sqrt{6}}{3}e^t\sin\sqrt{6}t, \quad y = 1 + e^t\cos\sqrt{6}t$$

(iii) Spirals away from (1, 2) to ∞

(d) (i) $\quad x = Ae^{5t} + Be^{-t}, \quad y = Ae^{5t} - Be^{-t}$

(ii)

$$x = \frac{1}{2}(3e^{5t} - e^{-t}), \quad y = \frac{1}{2}(3e^{5t} + e^{-t})$$

(iii) $(x, y) \to \infty$ as $t \to \infty$ along $y = x$

(e) (i) $\quad x = e^{-t}(A\sin t + B\cos t)$

$$y = \frac{e^{-t}}{5}[(A - 2B)\cos t - (2A + B)\sin t]$$

(ii) $x = e^{-t}(12\sin t + \cos t),$
$\quad\ y = e^{-t}(2\cos t - 5\sin t)$

(iii) $(x, y) \to (0, 0)$ as $t \to \infty$

(f) (i) $\quad x = e^{2t}(A + Bt) + 2, \quad y = -Be^{2t} + 3$

(ii) $\quad x = 2 + (t - 1)e^{2t}, \quad y = 3 - e^{2t}$

(iii) $x \ \rangle \ \infty, y \quad \rangle \quad \infty$ as $t \ \rangle \ \infty$

2. $\quad a > b + c, \quad \dfrac{T}{T + F} \to \dfrac{a - b - c}{a - c}$

3. (i) $\quad x = 8e^{-4t}$
$\quad\quad y = 16e^{-2t} - 16e^{-4t}$
$\quad\quad z = 8e^{-4t} - 16e^{-2t} + 8$
(ii) 4
(iii) $z \to 8$ as $t \to \infty$

4. (i) $\quad x = e^{-0.5t}$
(ii) $\quad y = 2.5(e^{-0.3t} - e^{-0.5t})$
(iii) $\quad z = 1.5e^{-0.5t} - 2.5e^{-0.3t} + 1$

5. (ii) $\quad x = e^{0.01t}(49\ 900t + 10\ 000)$
(iii) $\quad y = e^{0.01t}(49\ 900t + 5\ 000\ 000)$
(iv) $\quad x = 273\ 000$
$\quad\quad y = 5.52$ million

6. (ii) $\quad C_1 = 2.4 + 0.6e^{-0.05t}$
(iii) $\quad C_2 = 2.4 - 2.4e^{-0.05t}$
(iv) After 4 hours, $C_1 = 2.89 \quad C_2 = 0.435$

As $t \to \infty$, $C_1 \to \dfrac{12}{5} \quad C_2 \to \dfrac{12}{5}$

7. (i) $-5, -13$
(iii) $x = 5.75e^{-t} + 0.25e^{3t} - 2$
(iv) $y = 11.5e^{-t} - 0.5e^{3t} - 5$

8. (ii) $x = Ae^{-0.054t} + Be^{-0.026t} + 536$
(iii) $y = -0.4Ae^{-0.054t} + 2.4Be^{-0.026t} + 179$
(iv) $x = -395.5e^{-0.054t} - 140.5e^{-0.026t} + 536$
$\quad\quad y = 158.2e^{-0.054t} - 337.2e^{-0.026t} + 179$

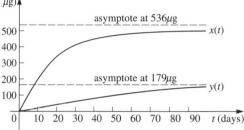

(v) $x = 536$ gm $\quad\quad y = 179$ gm

9. (i) $0, -20$
(ii) $\dfrac{d^2x}{dt^2} - 3\dfrac{dx}{dt} - 4x = 0$

(iii) $x = 3e^{4t} + 12e^{-t}$
(iv) $y = 12e^{-t} - 2e^{4t}$
(v) The second species becomes extinct after nearly 36 years.

10. $\dfrac{dp}{dt} = kp \quad \dfrac{dq}{dt} = 2kq \quad$ 45 grams

11.

$$10^4\dfrac{d^2x}{dt^2} - 800\dfrac{dx}{dt} + 16x = 300\cos t + 10000\sin t$$

$$x = (At + B)e^{0.04t} + 0.05\cos t - 1.00\sin t$$

12. $q = 2e^{-t} - 2e^{-2t}$
13. (a) $(1.5, -2.5)$
(b) $(-1, 1)$
(c) $(-1, -1)$ $(-2, -2)$
14. (i) All are 2.4×10^{-10} kgl^{-1}
(ii) 16.2 years
(iii) 16.2 years
15. (ii) $y = 90.91$
$\quad\quad x = 181.82$
(iii) A : x increases from 0 to 181.82
$\quad$ y initially decreases to a minimum and then increases to 90.91
$\quad$ B : Both x and y decrease to their equilibrium values.
16. (i) (a) $P = 25.8, \quad Q = 774.6$
17. Equilibrium values for (F, R)
$\quad$ $(0, 1200)$ $(-40, 0)$, $(10, 1000)$ and $(0, 0)$.
$\quad$ Only $(10, 1000)$ is sensible since $R, F > 0$.

Index